Section 8

Process Control

Thomas F. Edgar, Ph.D. *Professor of Chemical Engineering, University of Texas—Austin (Section Editor, Advanced Control Systems, Process Measurements)*

Cecil L. Smith, Ph.D. *Principal, Cecil L. Smith Inc. (Batch Process Control, Telemetering and Transmission, Digital Technology for Process Control, Process Control and Plant Safety)*

F. Greg Shinskey, B.S.Ch.E. *Consultant (retired from Foxboro Co.) (Fundamentals of Process Dynamics and Control, Unit Operations Control)*

George W. Gassman, B.S.M.E. *Senior Research Specialist, Final Control Systems, Fisher Controls International, Inc. (Controllers, Final Control Elements, and Regulators)*

Andrew W. R. Waite, P.Eng. *Principal Process Control Consultant, EnTech Control, a Division of Emerson Electric Canada (Controllers, Final Control Elements, and Regulators)*

Thomas J. McAvoy, Ph.D. *Professor of Chemical Engineering, University of Maryland—College Park (Fundamentals of Process Dynamics and Control)*

Dale E. Seborg, Ph.D. *Professor of Chemical Engineering, University of California—Santa Barbara (Advanced Control Systems)*

FUNDAMENTALS OF PROCESS DYNAMICS AND CONTROL

ADVANCED CONTROL SYSTEMS

UNIT OPERATIONS CONTROL

BATCH PROCESS CONTROL

PROCESS MEASUREMENTS

TELEMETERING AND TRANSMISSION

DIGITAL TECHNOLOGY FOR PROCESS CONTROL

CONTROLLERS, FINAL CONTROL ELEMENTS, AND REGULATORS

PROCESS CONTROL AND PLANT SAFETY

Nomenclature

Symbol	Definition
A	Area
A_a	Actuator area
A_c	Output amplitude limits
A_v	Amplitude of controlled variable
A_1	Cross-sectional area of tank
b	Controller output bias
B	Bottoms flow rate
B_i°	Limit on control
c_A	Concentration of A
C	Cumulative sum
C_d	Discharge coefficient
C_i	Inlet concentration
C_i°	Limit on control move
C_L	Specific heat of liquid
C_0	Integration constant
C_p	Process capability
C_r	Heat capacity of reactants
C_V	Valve flow coefficient
D	Distillate flow rate, disturbance
D_i°	Limit on output
e	Error
E	Economy of evaporator
f	Function of time
F, f	Feed flow rate
F_L	Pressure recovery factor
g_c	Unit conversion constant
g_i	Algebraic inequality constraint
G	Transfer function
G_c	Controller transfer function
G_d	Disturbance transfer function
G_f	Feedforward controller transfer function
G_m	Sensor transfer function
G_p	Process transfer function
G_t	Transmitter transfer function
G_v	Valve transfer function
h_i	Algebraic equality constraints
h_1	Liquid head in tank
H	Latent heat of vaporization, control limit or threshold
i	Summation index
I_i	Impulse response coefficient
j	Time index
J	Objective function or performance index
k	Time index
k_f	Flow coefficient
k_r	Kinetic rate constant
K	Gain, slack parameter
K_c	Controller gain
K_d	Disturbance transfer function gain
K_m	Measurement gain
K_p	Process gain
K_u	Ultimate controller gain (stability)
L	Load variable
L_p	Sound pressure level
M	Manipulated variable
m_c	Number of constraints
M_v	Mass flow
M_r	Mass of reactants
M_w	Molecular weight
n	Number of data points, number of stages or effects
N	Number of inputs/outputs, model horizon
p	Proportional band (%)
p_c	Vapor pressure
p_d	Actuator pressure
p_i	Pressure
p_u	Proportional band (ultimate)
q	Radiated energy flux
q_b	Energy flux to a black body
Q	Flow rate
r_c	Number of constraints
R	Equal-percentage valve characteristic
R_T	Resistance in temperature sensor
R_1	Valve resistance

Symbol	Definition
s	Laplace transform variable
s	Search direction
S_i	Step response coefficient
t	Time
T	Temperature, target
$T(s)$	Decoupler transfer function
T_b	Base temperature
T_f	Exhaust temperature
T_R	Reset time
U	Heat-transfer coefficient
u, U	Manipulated variable, controller output
V	Volume
V_s	Product value
w	Mass flow rate
w_i	Weighting factor
W	Steam flow rate
x	Mass fraction
$\overline{x}$	Sample mean
x_i	Optimization variable
x_T	Pressure drop ratio factor
X	Transform of deviation variable
y, Y	Process output, controlled variable, valve travel
Y_{sp}	Set point
z	Controller tuning law, expansion factor
z_i	Feed mole fraction (distillation)
Z	Compressibility factor

Greek Symbols

Symbol	Definition
α	Digital filter coefficient
a_T	Temperature coefficient of resistance
β	Resistance thermometer parameter
γ	Ratio of specific heats
δ	Move suppression factor, shift in target value
Δq	Load step change
Δt	Time step
ΔT	Temperature change
Δu	Control move
ε	Spectral emissivity, step size
ζ	Damping factor (second-order system)
θ	Time delay
λ	Relative gain array parameter, wavelength
Λ	Relative gain array
ξ	Deviation variable
ρ	Density
σ	Stefan-Boltzmann constant, standard deviation
Σ_t	Total response time
τ	Time constant
τ_d	Natural period of closed loop, disturbance time constant
τ_D	Derivative time (PID controller)
τ_F	Filter time constant
τ_I	Integral time (PID controller)
τ_P	Process time constant
τ_o	Period of oscillation
ϕ_{PI}	Phase lag

Subscripts

Symbol	Definition
A	Species A
b	Best
c	Controller
d	Disturbance
eff	Effective
f	Feedforward
i	Initial, inlet
L	Load, disturbance
m	Measurement or sensor
p	Process
s	Steady state
sp	Set-point value
t	Transmitter
u	Ultimate
v	Valve

FUNDAMENTALS OF PROCESS DYNAMICS AND CONTROL

THE GENERAL CONTROL SYSTEM

A process is shown in Fig. 8-1 with a manipulated input U, a load input D, and a controlled output Y, which could be flow, pressure, liquid level, temperature, composition, or any other inventory, environmental, or quality variable that is to be held at a desired value identified as the set point Y_{sp}. The load may be a single variable or an aggregate of variables either acting independently or manipulated for other purposes, affecting the controlled variable much as the manipulated variable does. Changes in load may occur randomly as caused by changes in weather, diurnally with ambient temperature, manually when operators change production rate, stepwise when equipment is switched into or out of service, or cyclically as the result of oscillations in other control loops. Variations in load will drive the controlled variable away from the set point, requiring a corresponding change in the manipulated variable to bring it back. The manipulated variable must also change to move the controlled variable from one set point to another.

An open-loop system positions the manipulated variable either manually or on a programmed basis, without using any process measurements. This operation is acceptable for well-defined processes without disturbances. An automated transfer switch is provided to allow manual adjustment of the manipulated variable in case the process or the control system is not performing satisfactorily.

A closed-loop system uses the measurement of one or more process variables to move the manipulated variable to achieve control. Closed-loop systems may include feedforward, feedback, or both.

Feedback Control In a feedback control loop, the controlled variable is compared to the set point Y_{sp}, with the error E acted upon by the controller to move U in such a way as to minimize the error. This action is specifically negative feedback, in that an increase in error moves U so as to decrease the error. (Positive feedback would cause the error to expand rather than diminish and therefore does not regulate.) The action of the controller is selectable to allow use on process gains of both signs.

The controller has tuning parameters related to proportional, integral, derivative, lag, dead time, and sampling functions. A negative feedback loop will oscillate if the controller gain is too high; but if it is too low, control will be ineffective. The controller parameters must be properly related to the process parameters to ensure closed-loop stability while still providing effective control. This relationship is accomplished, first, by the proper selection of control modes to satisfy the requirements of the process and, second, by the appropriate tuning of those modes.

Feedforward Control A feedforward system uses measurements of disturbance variables to position the manipulated variable in such a way as to minimize any resulting deviation. The disturbance variables could be either measured loads or the set point, the former being more common. The feedforward gain must be set precisely to reduce the deviation of the controlled variable from the set point.

Feedforward control is usually combined with feedback control to eliminate any offset resulting from inaccurate measurements and calculations and unmeasured load components. The feedback controller can be used as a bias on the feedforward controller or in a multiplicative form.

Computer Control Computers have been used to replace analog PID controllers, either by setting set points of lower-level controllers in supervisory control or by driving valves directly in direct digital control. Single-station digital controllers perform PID control in one or two loops, including computing functions such as mathematical operations, characterization, lags, and dead time, with digital logic and alarms. Distributed control systems provide all these functions, with the digital processor shared among many control loops; separate processors may be used for displays, communications, file servers, and the like. A host computer may be added to perform high-level operations such as scheduling, optimization, and multivariable control. More details on computer control are provided later in this section.

PROCESS DYNAMICS AND MATHEMATICAL MODELS

General References: Seborg, Edgar, and Mellichamp, *Process Dynamics and Control,* Wiley, New York, 2004; Marlin, *Process Control,* McGraw-Hill, New York, 2000; Ogunnaike and Ray, *Process Dynamics Modeling and Control,* Oxford University Press, New York, 1994; Smith and Corripio, *Principles and Practices of Automatic Process Control,* Wiley, New York, 1997.

Open-Loop versus Closed-Loop Dynamics It is common in industry to manipulate coolant in a jacketed reactor in order to control conditions in the reactor itself. A simplified schematic diagram of such a reactor control system is shown in Fig. 8-2. Assume that the reactor temperature is adjusted by a controller that increases the coolant flow in proportion to the difference between the desired reactor temperature and the temperature that is measured. The proportionality constant is K_c. If a small change in the temperature of the inlet stream occurs, then

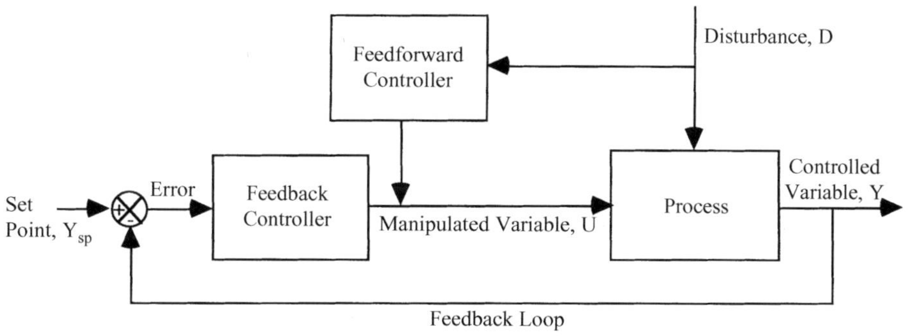

FIG. 8-1 Block diagram for feedforward and feedback control.

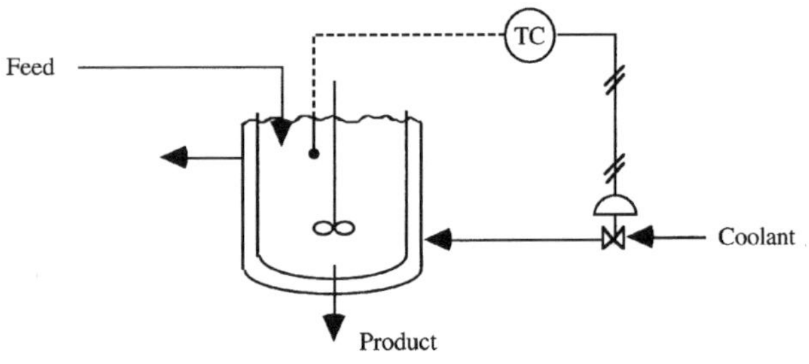

FIG. 8-2 Reactor control system.

depending on the value of K_c, one might observe the reactor temperature responses shown in Fig. 8-3. The top plot shows the case for no control ($K_c = 0$), which is called the open loop, or the normal dynamic response of the process by itself. As K_c increases, several effects can be noted. First, the reactor temperature responds faster and faster. Second, for the initial increases in K, the maximum deviation in the reactor temperature becomes smaller. Both of these effects are desirable so that disturbances from normal operation have as small an effect as possible on the process under study. As the gain is increased further, eventually a point is reached where the reactor temperature oscillates indefinitely, which is undesirable. This point is called the stability limit, where $K_c = K_u$, the ultimate controller gain. Increasing K_c further causes the magnitude of the oscillations to increase, with the result that the control valve will cycle between full open and closed.

The responses shown in Fig. 8-3 are typical of the vast majority of regulatory loops encountered in the process industries. Figure 8-3 shows that there is an optimal choice for K_c, somewhere between 0 (no control) and K_u (stability limit). If one has a dynamic model of a process, then this model can be used to calculate controller settings. In Fig. 8-3, no time scale is given, but rather the figure shows relative responses. A well-designed controller might be able to speed up the response of a process by a factor of roughly 2 to 4. Exactly how fast the control system responds is determined by the dynamics of the process itself.

Physical Models versus Empirical Models In developing a dynamic process model, there are two distinct approaches that can be taken. The first involves models based on first principles, called physical or first principles models, and the second involves empirical models. The conservation laws of mass, energy, and momentum form the basis for developing physical models. The resulting models typically involve sets of differential and algebraic equations that must be solved simultaneously. Empirical models, by contrast, involve postulating the form of a dynamic model, usually as a transfer function, which is discussed below. This transfer function contains a number of parameters that need to be estimated from data. For the development of both physical and empirical models, the most expensive step normally involves verification of their accuracy in predicting plant behavior.

To illustrate the development of a physical model, a simplified treatment of the reactor, shown in Fig. 8-2, is used. It is assumed that the reactor is operating isothermally and that the inlet and exit volumetric flows and densities are the same. There are two components, A and B, in the reactor, and a single first-order reaction of $A \rightarrow B$ takes place. The inlet concentration of A, which we call c_i, varies with time. A dynamic mass balance for the concentration of A, denoted c_A, can be written as follows:

$$V\frac{dc_A}{dt} = Fc_i - Fc_A - k_r V c_A \tag{8-1}$$

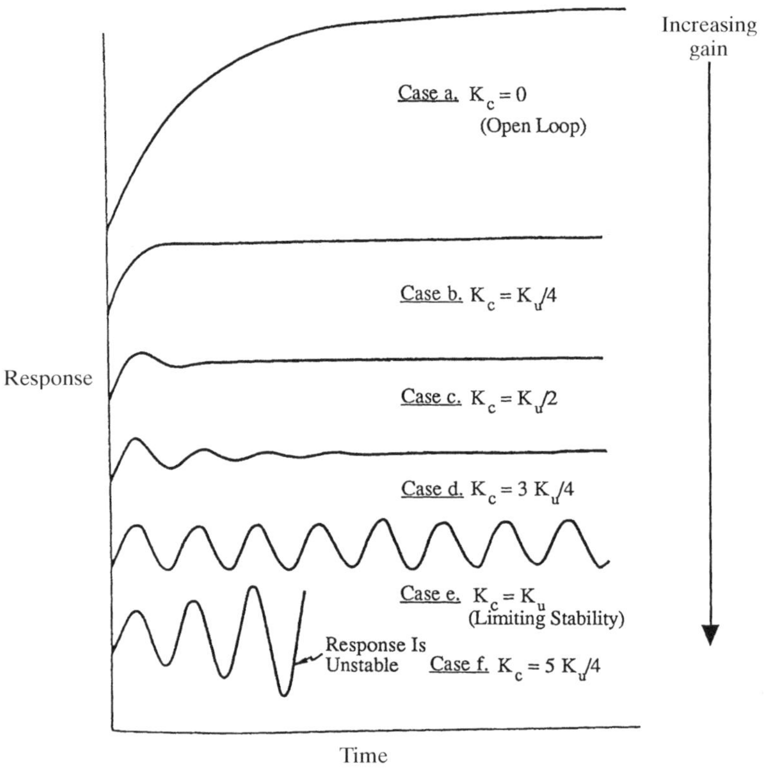

FIG. 8-3 Typical control system responses.

In Eq. (8-1), the flow in of component A is Fc_i, the flow out is Fc_A, and the loss via reaction is k_rVc_A, where V = reactor volume and k_r = kinetic rate constant. In this example, c_i is the input, or forcing, variable and c_A is the output variable. If V, F, and k_r are constant, Eq. (8-1) can be rearranged by dividing by $F + k_rV$ so that it contains only two groups of parameters. The result is

$$\tau \frac{dc_A}{dt} = Kc_i - c_A \tag{8-2}$$

where $\tau = V/(F + k_rV)$ and $K = F/(F + k_rV)$. For this example, the resulting model is a first-order differential equation in which τ is called the time constant and K the process gain.

As an alternative to deriving Eq. (8-2) from a dynamic mass balance, one could simply postulate a first-order differential equation to be valid (empirical modeling). Then it would be necessary to estimate values for τ and K so that the postulated model described the reactor's dynamic response. The advantage of the physical model over the empirical model is that the physical model gives insight into how reactor parameters affect the values of τ and K, which in turn affects the dynamic response of the reactor.

Nonlinear versus Linear Models If V, F, and k are constant, then Eq. (8-1) is an example of a linear differential equation model. In a linear equation, the output and input variables and their derivatives appear to only the first power. If the rate of reaction were second-order, then the resulting dynamic mass balance would be

$$V\frac{dc_A}{dt} = Fc_i - Fc_A - k_rVc_A^2 \tag{8-3}$$

Since c_A appears in this equation to the second power, the equation is nonlinear.

The difference between linear systems and nonlinear systems can be seen by considering the steady-state behavior of Eq. (8-1) compared to Eq. (8-3) (the left-hand side is zero; that is, $dc_A/dt = 0$). For a given change in c_i, Δc_i, the change in c_A calculated from Eq. (8-1), Δc_A, is always proportional to Δc_i, and the proportionality constant is K [see Eq. (8-2)]. The change in the output of a system divided by a change in the input to the system is called the *process gain*. Linear systems have constant process gains for all changes in the input. By contrast, Eq. (8-3) gives a Δc_A that varies with Δc_i, which is a function of the concentration levels in the reactor. Thus, depending on the reactor operating conditions, a change in c_i produces different changes in c_A. In this case, the process has a nonlinear gain. Systems with nonlinear gains are more difficult to control than linear systems that have constant gains.

Simulation of Dynamic Models Linear dynamic models are particularly useful for analyzing control system behavior. The insight gained through linear analysis is invaluable. However, accurate dynamic process models can involve large sets of nonlinear equations. Analytical solution of these models is not possible. Thus, in these cases, one must turn to simulation approaches to study process dynamics and the effect of process control. Equation (8-3) will be used to illustrate the simulation of nonlinear processes. If dc_A/dt on the left-hand side of Eq. (8-3) is replaced with its finite difference approximation, one gets

$$c_A(t + \Delta t) = \frac{c_A(t) + \Delta t \cdot [Fc_i(t) - Fc_A(t) - k_rVc_A^2(t)]}{V} \tag{8-4}$$

Starting with an initial value of c_A and given $c_i(t)$, Eq. (8-4) can be solved for $c_A(t + \Delta t)$. Once $c_A(t + \Delta t)$ is known, the solution process can be repeated to calculate $c_A(t + 2\Delta t)$, and so on. This approach is called the Euler integration method; while it is simple, it is not necessarily the best approach to numerically integrating nonlinear differential equations. As discussed in Sec. 3, more sophisticated approaches are available that allow much larger step sizes to be taken but require additional calculations. One widely used approach is the fourth-order Runge Kutta method, which involves the following calculations:

Define

$$f(c_At) = \frac{Fc_i(t) - Fc_A - k_rVc_A^2}{V} \tag{8-5}$$

then

$$c_A(t + \Delta t) = c_A(t) + \Delta t(m_1 + 2m_2 + 2m_3 + m_4) \tag{8-6}$$

with

$$m_1 = f[c_A(t), t] \tag{8-7}$$

$$m_2 = f\left[c_A(t) + 2\frac{m_1\,\Delta t}{2}, t + \frac{\Delta t}{2}\right] \tag{8-8}$$

$$m_3 = f\left[c_A(t) + 2\frac{m_2\,\Delta t}{2}, t + \frac{\Delta t}{2}\right] \tag{8-9}$$

$$m_4 = f[c_A(t) + m_3\,\Delta t, t + \Delta t] \tag{8-10}$$

In this method, the m_i's are calculated sequentially in order to take a step in time. Even though this method requires calculation of the four additional m_i values, for equivalent accuracy the fourth-order Runge Kutta method can result in a faster numerical solution, because it permits a larger step Δt to be taken. Increasingly sophisticated simulation packages are being used to calculate the dynamic behavior of processes and to test control system behavior. These packages have good user interfaces, and they can handle stiff systems where some variables respond on a time scale that is much much faster or slower than that of other variables. A simple Euler approach cannot effectively handle stiff systems, which frequently occur in chemical process models. See Sec. 3 of this handbook for more details.

Laplace Transforms When mathematical models are used to describe process dynamics in conjunction with control system analysis, the models generally involve linear differential equations. Laplace transforms are very effective for solving linear differential equations. The key advantage of using Laplace transforms is that they convert differential equations to algebraic equations. The resulting algebraic equations are easier to solve than the original differential equations. When the Laplace transform is applied to a linear differential equation in time, the result is an algebraic equation in a new variable s, called the Laplace variable. To get the solution to the original differential equation, one needs to invert the Laplace transform. Table 8-1 gives a number of useful Laplace transform pairs, and more extensive tables are available (Seborg, Edgar, and Mellichamp, *Process Dynamics and Control*, Wiley, New York, 2004).

To illustrate how Laplace transforms work, consider the problem of solving Eq. (8-2), subject to the initial condition that $c_A = c_i = 0$ at $t = 0$. If c_A were not initially zero, one would define a deviation variable between c_A and its initial value c_{A0}. Then the transfer function would be developed by using this deviation variable. If c_i changes from zero to $\bar{c}_i$, taking the Laplace transform of both sides of Eq. (8-2) gives

$$\pounds\left(\frac{\tau dc_A}{dt}\right) = \pounds(K\bar{c}_i) - \pounds(c_A) \tag{8-11}$$

TABLE 8-1 Frequently Used Laplace Transforms

Time function $f(t)$	Transform $F(s)$
A	A/s
At	A/s^2
Ae^{-at}	$A/(s + a)$
$A(1 - e^{-t/\tau})$	$A/[s(\tau s + 1)]$
$A \sin \omega t$	$A\omega/(s^2 + \omega^2)$
$f(t - \theta)$	$e^{-\theta}F(s)$
df/dt	$sF(s) - f(0)$
$\int f(t)dt$	$F(s)/s$

Denoting $\pounds(C_A)$ as $C_A(s)$ and using the relationships in Table 8-1 give

$$\tau s C_A(s) = \frac{K\bar{c}_i}{s} - C_A(s) \tag{8-12}$$

Equation (8-12) can be solved for C_A to give

$$C_A(s) = \frac{K\bar{c}_i/s}{\tau s + 1} \tag{8-13}$$

By using the entries in Table 8-1, Eq. (8-13) can be inverted to give the transient response of c_A as

$$c_A(t) = (K\bar{c}_i)(1 - e^{-1/\tau}) \tag{8-14}$$

Equation (8-14) shows that c_A starts from 0 and builds up exponentially to a final concentration of Kc_i. Note that to get Eq. (8-14), it was only necessary to solve the algebraic Eq. (8-12) and then find the inverse of $C_A(s)$ in Table 8-1. The original differential equation was not solved directly. In general, techniques such as partial fraction expansion must be used to solve higher-order differential equations with Laplace transforms.

Transfer Functions and Block Diagrams A very convenient and compact method of representing the process dynamics of linear systems involves the use of transfer functions and block diagrams. A transfer function can be obtained by starting with a physical model, as discussed previously. If the physical model is nonlinear, first it needs to be linearized around an operating point. The resulting linearized model is then approximately valid in a region around this operating point. To illustrate how transfer functions are developed, Eq. (8-2) will again be used. First, one defines deviation variables, which are the process variables minus their steady-state values at the operating point. For Eq. (8-2), there would be deviation variables for both c_A and c_i, and these are defined as

$$\xi = c_A - \bar{c}_A \tag{8-15}$$

$$\xi_i = c_i - \bar{c}_i \tag{8-16}$$

where the overbar stands for steady state. Substitution of Eqs. (8-15) and (8-16) into Eq. (8-2) gives

$$\tau \frac{d\xi}{dt} = K\xi_i - \xi + (K\bar{c}_i - \bar{c}_A) \tag{8-17}$$

The term in parentheses in Eq. (8-17) is zero at steady state, and thus it can be dropped. Next the Laplace transform is taken, and the resulting algebraic equation is solved.

By denoting $X(s)$ as the Laplace transform of ξ and $X_i(s)$ as the transform of ξ_i, the final transfer function can be written as

$$\frac{X}{X_i} = \frac{K}{\tau s + 1} \tag{8-18}$$

Equation (8-18) is an example of a first-order transfer function. As mentioned above, an alternative to formally deriving Eq. (8-18) involves simply postulating its form and then identifying its two parameters, the process gain K and time constant τ, to fit the process under study. In fitting the parameters, data can be generated by forcing the process. If step forcing is used, then the resulting response is called the process reaction curve. Often transfer functions are placed in block diagrams, as shown in Fig. 8-4. Block diagrams show how changes in an input variable affect an output variable. Block diagrams are a means of concisely representing the dynamics of a process under study. Since linearity is assumed in developing a block diagram, if more than one variable affects an output, the contributions from each can be added.

Continuous versus Discrete Models The preceding discussion has focused on systems where variables change continuously with

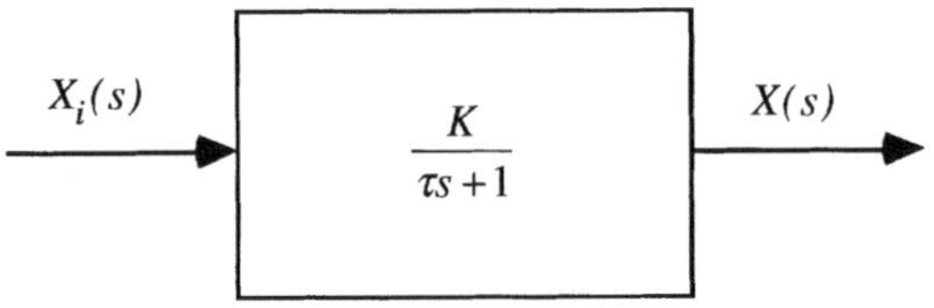

FIG. 8-4 First-order transfer function.

time. Most real processes have variables that are continuous, such as temperature, pressure, and flow. However, some processes involve discrete events, such as the starting or stopping of a pump. In addition, modern plants are controlled by digital computers, which are discrete. In controlling a process, a digital system samples variables at a fixed rate, and the resulting system is a sampled data system. From one sampling instant until the next, variables are assumed to remain fixed at their sampled values. Similarly, in controlling a process, a digital computer sends out signals to control elements, usually valves, at discrete instants of time. These signals remain fixed until the next sampling instant.

Figure 8-5 illustrates the concept of sampling a continuous function. At integer values of the sampling rate Δt, the value of the variable to be sampled is measured and held until the next sampling instant. To deal with sampled data systems, the z transform has been developed. The z transform of the function given in Fig. 8-5 is defined as

$$Z(f) = \sum_{n=0}^{\infty} f(n\ \Delta t) z^{-n} \tag{8-19}$$

In an analogous manner to Laplace transforms, one can develop transfer functions in the z domain as well as block diagrams. Tables of z transform pairs have been published (Seborg, Edgar, and Mellichamp, *Process Dynamics and Control,* Wiley, New York, 2004) so that the discrete transfer functions can be inverted back to the time domain. The inverse gives the value of the function at the discrete sampling instants. Sampling a continuous variable results in a loss of information. However, in practical applications, sampling is fast enough that the loss is typically insignificant and the difference between continuous and discrete modeling is small in terms of its effect on control. Increasingly, model predictive controllers that make use of discrete dynamic models are being used in the process industries. The purpose of these controllers is to guide a process to optimum operating points. These model predictive control algorithms are typically run at much slower sampling rates than are used for basic control loops such as flow control or pressure control. The discrete dynamic models used are normally developed from data generated

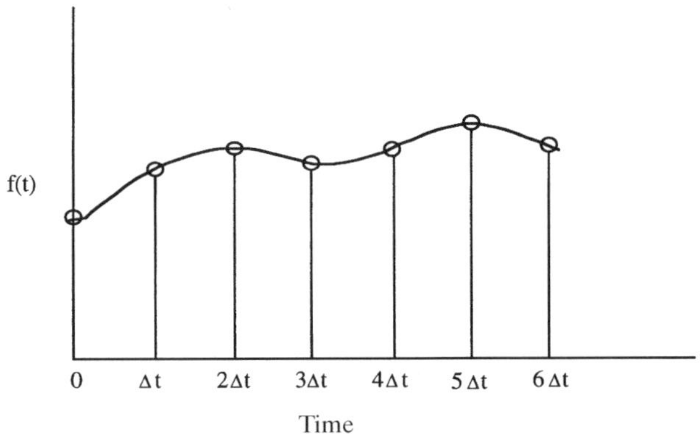

FIG. 8-5 Sampled data example.

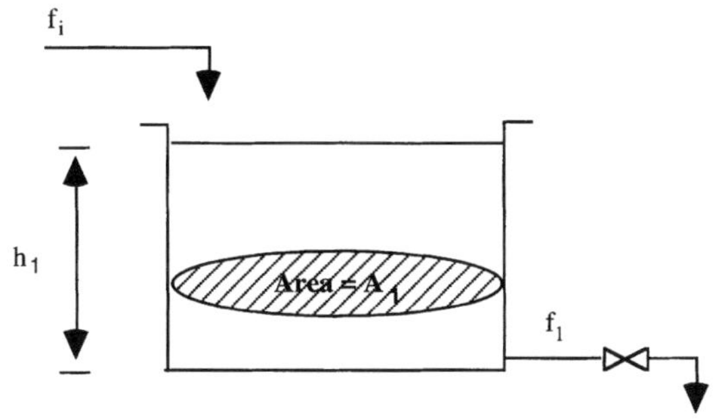

FIG. 8-6 Single tank with exit valve.

from plant testing, as discussed hereafter. For a detailed discussion of modeling sampled data systems, the interested reader is referred to textbooks on digital control (Astrom and Wittenmark, *Computer Controlled Systems*, Prentice-Hall, Englewood Cliffs, N.J., 1997).

Process Characteristics in Transfer Functions In many cases, process characteristics are expressed in the form of transfer functions. In the previous discussion, a reactor example was used to illustrate how a transfer function could be derived. Here, another system involving flow out of a tank, shown in Fig. 8-6, is considered.

Proportional Element First, consider the outflow through the exit valve on the tank. If the flow through the line is turbulent, then Bernoulli's equation can be used to relate the flow rate through the valve to the pressure drop across the valve as

$$f_1 = k_f A_v \sqrt{2g_c(h_1 - h_0)} \tag{8-20}$$

where f_1 = flow rate, k_f = flow coefficient, A_v = cross-sectional area of the restriction, g_c = constant, h_1 = liquid head in tank (pressure at the base of the tank), and h_0 = atmospheric pressure. This relationship between flow and pressure drop across the valve is nonlinear, and it can be linearized around a particular operating point to give

$$f_1 - \bar{f}_1 = \left(\frac{1}{R_1}\right)(h_1 - \bar{h}_1) \tag{8-21}$$

where $R_1 = \bar{f}_1/(g_c k_f^2 A_v^2)$ is called the resistance of the valve in analogy with an electrical resistance. The transfer function relating changes in flow to changes in head is shown in Fig. 8-7, and it is an example of a pure gain system with no dynamics. In this case, the process gain is $K = 1/R_1$. Such a system has an instantaneous dynamic response, and for a step change in head, there is an immediate step change in flow, as shown in Fig. 8-8. The exact magnitude of the step in flow depends on the operating flow $\bar{f}_1$ as the definition of R_1 shows.

First-Order Lag (Time Constant Element) Next consider the system to be the tank itself. A dynamic mass balance on the tank gives

$$A_1 \frac{dh_1}{dt} = f_i - f_1 \tag{8-22}$$

where A_1 is the cross-sectional area of the tank and f_i is the inlet flow. By substituting Eq. (8-21) into Eq. (8-22) and following the approach

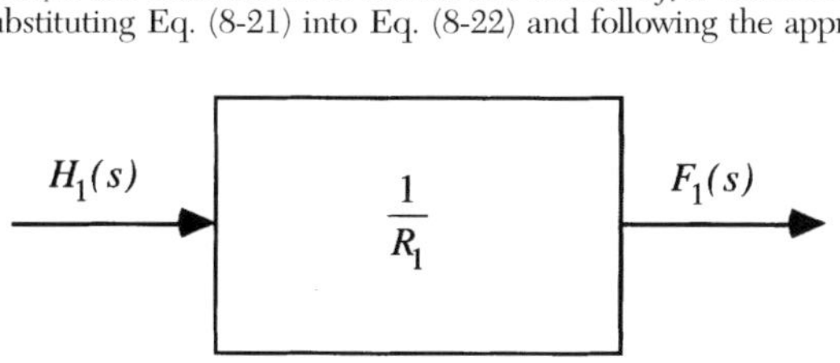

FIG. 8-7 Proportional element transfer function.

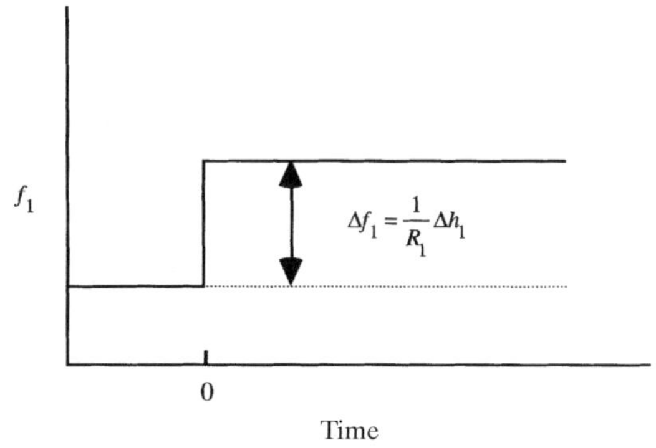

FIG. 8-8 Response of proportional element.

discussed above for deriving transfer functions, one can develop the transfer function relating changes in h_1 to changes in f_i. The resulting transfer function is another example of a first-order system, shown in Fig. 8-4, and it has a gain $K = R_1$ and a time constant $\tau_1 = R_1A_1$. For a step change in f_i, h_1 follows a decaying exponential response from its initial value $\bar{h}_1$ to a final value of $\bar{h}_1 + R_1\,\Delta f_i$ (Fig. 8-9). At a time equal to τ_1, the transient in h_1 is 63 percent finished; and at $3\tau_1$, the response is 95 percent finished. These percentages are the same for all first-order processes. Thus, knowledge of the time constant of a first-order process gives insight into how fast the process responds to sudden input changes.

Capacity Element Now consider the case where the valve in Fig. 8-7 is replaced with a pump. In this case, it is reasonable to assume that the exit flow from the tank is independent of the level in the tank. For such a case, Eq. (8-22) still holds, except that f_1 no longer depends on h_1. For changes in f_i, the transfer function relating changes in h_1 to changes in f_i is shown in Fig. 8-10. This is an example of a pure capacity process, also called an integrating system. The cross-sectional area of the tank is the chemical process equivalent of an electrical capacitor. If the inlet flow is step forced while the outlet is held constant, then the level builds up linearly, as shown in Fig. 8-11. Eventually the liquid will overflow the tank.

Second-Order Element Because of their linear nature, transfer functions can be combined in a straightforward manner. Consider the two-tank system shown in Fig. 8-12. For tank 1, the transfer function relating changes in f_1 to changes in f_i is

$$\frac{F_1(s)}{F_i(s)} = \frac{1}{A_1R_1 + 1} \tag{8-23}$$

Since f_1 is the inlet flow to tank 2, the transfer function relating changes in h_2 to changes in f_1 has the same form as that given in Fig. 8-4:

$$\frac{H_2(s)}{F_1(s)} = \frac{R_2}{A_2R_2s + 1} \tag{8-24}$$

Equations (8-23) and (8-24) can be multiplied to give the final transfer function relating changes in h_2 to changes in f_i, as shown in Fig. 8-13. This is an example of a second-order transfer function. This transfer function has a gain R_2 and two time constants A_1R_1 and A_2R_2. For two tanks with equal areas, a step change in f_i produces the S-shaped response in level in the second tank shown in Fig. 8-14.

General Second-Order Element Figure 8-3 illustrates the fact that closed-loop systems can exhibit oscillatory behavior. A general second-order transfer function that can exhibit oscillatory behavior is important for the study of automatic control systems. Such a transfer function is given in Fig. 8-15. For a unit step input, the transient responses shown in Fig. 8-16 result. As can be seen, when $\zeta < 1$, the response oscillates; and when $\zeta < 1$, the response is S-shaped. Few open-loop chemical processes exhibit an oscillating response; most exhibit an S-shaped step response.

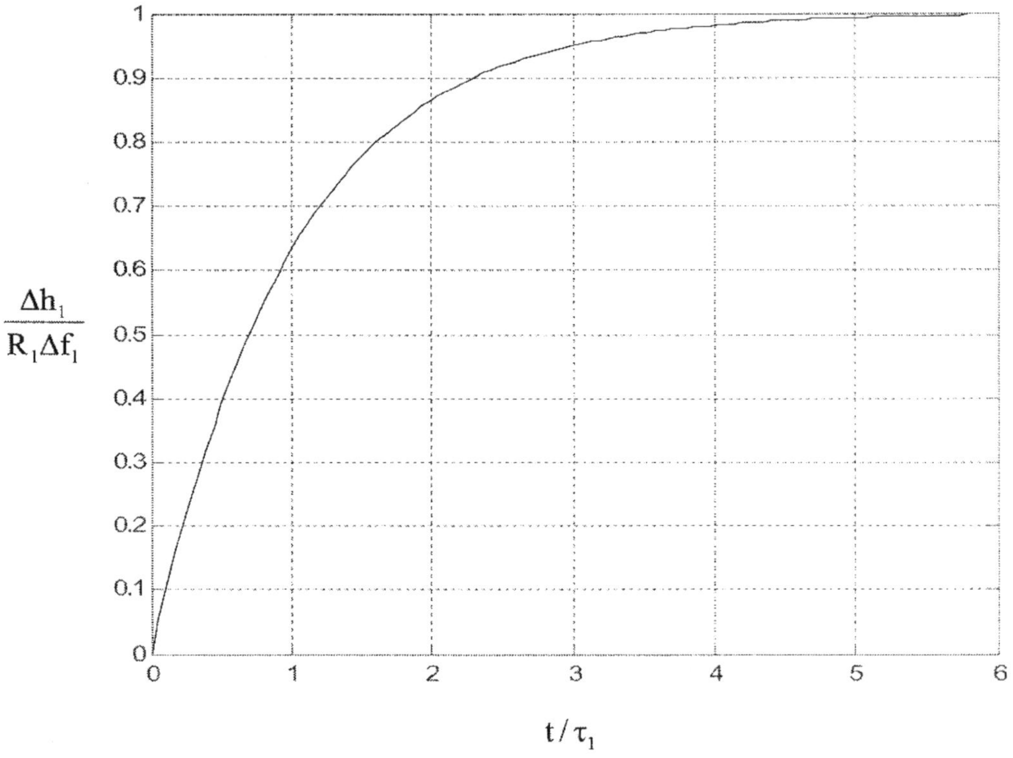

FIG. 8-9 Response of first-order system.

Distance-Velocity Lag (Dead-Time Element) The dead-time or time-delay element, commonly called a distance-velocity lag, is often encountered in process systems. For example, if a temperature-measuring element is located downstream from a heat exchanger, a time delay occurs before the heated fluid leaving the exchanger arrives at the temperature measurement point. If some element of a system produces a dead time of θ time units, then an input to that unit $f(t)$ will be reproduced at the output as $f(t - \theta)$. The transfer function for a pure dead-time element is shown in Fig. 8-17, and the transient response of the element is shown in Fig. 8-18.

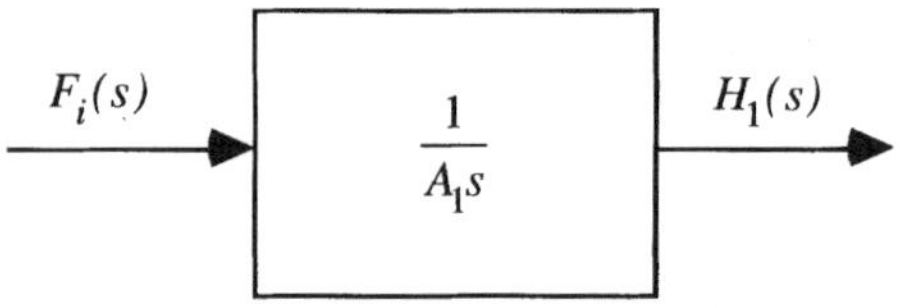

FIG. 8-10 Pure capacity or integrating transfer function.

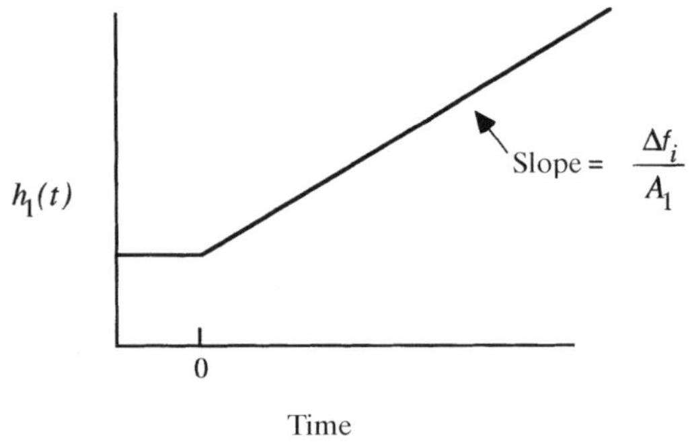

FIG. 8-11 Response of pure capacity system.

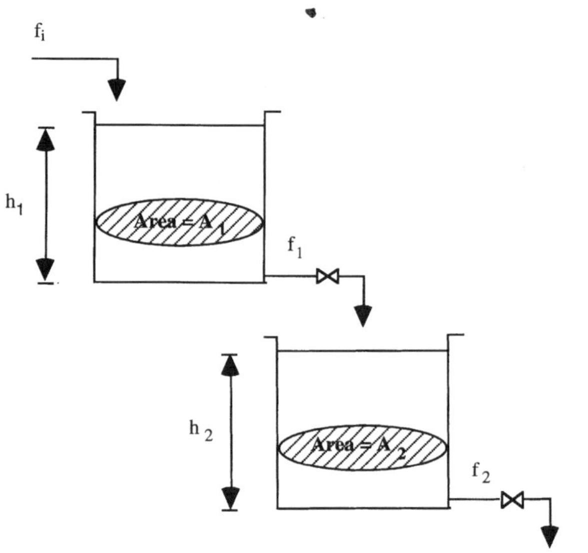

FIG. 8-12 Two tanks in series.

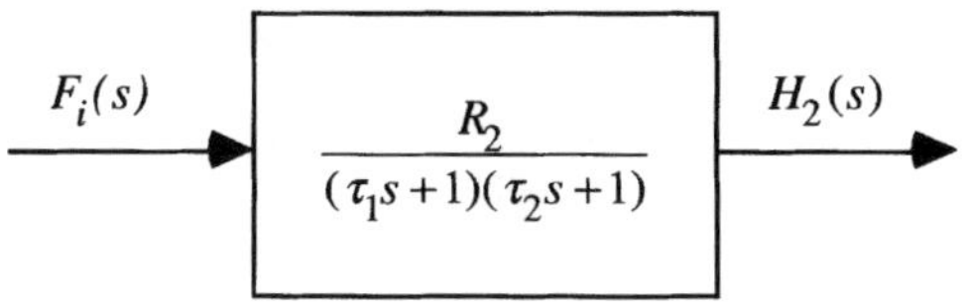

FIG. 8-13 Second-order transfer function.

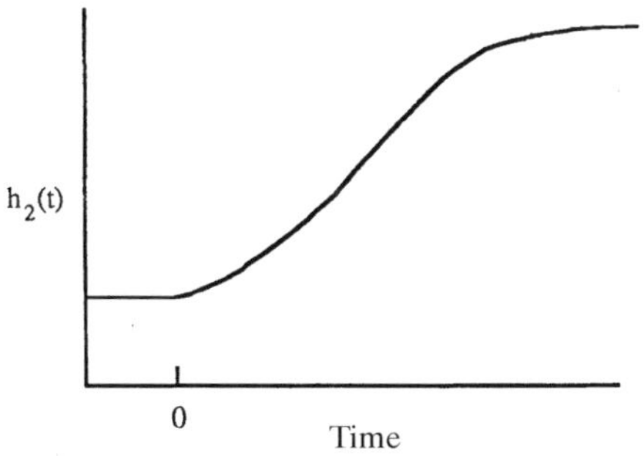

FIG. 8-14 Response of second-order system.

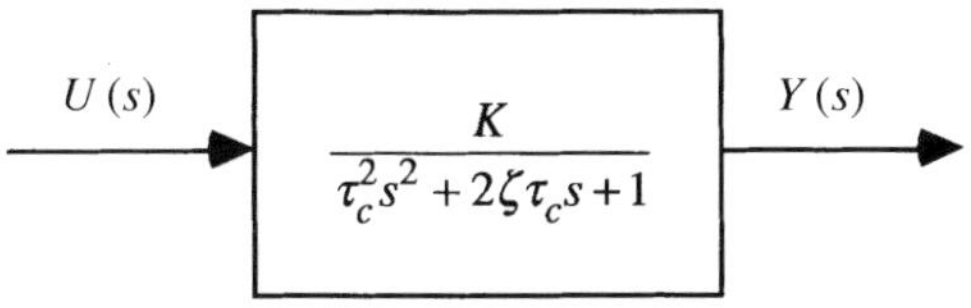

FIG. 8-15 General second-order transfer function.

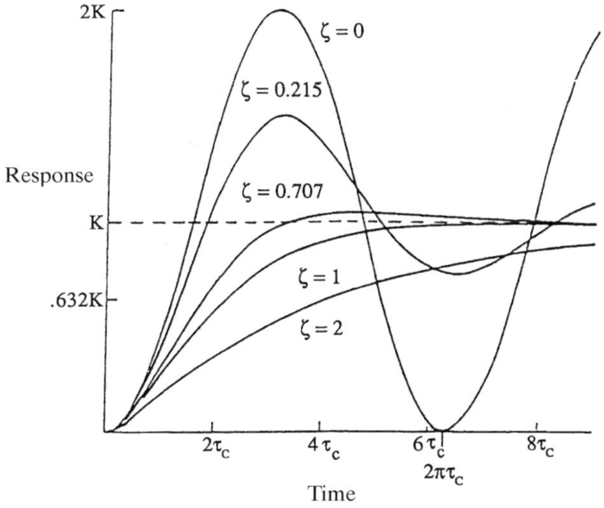

FIG. 8-16 Response of general second-order system.

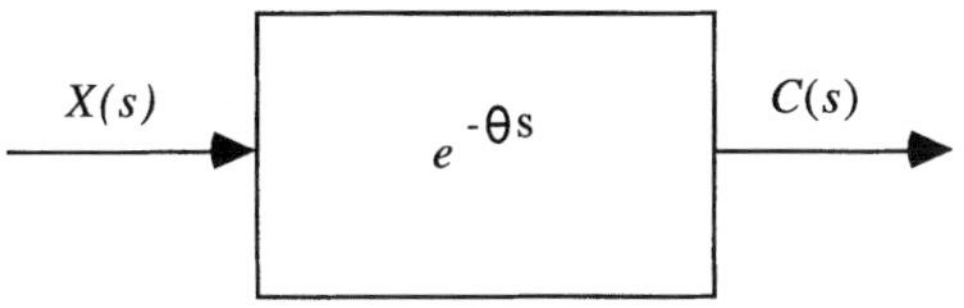

FIG. 8-17 Dead-time transfer function.

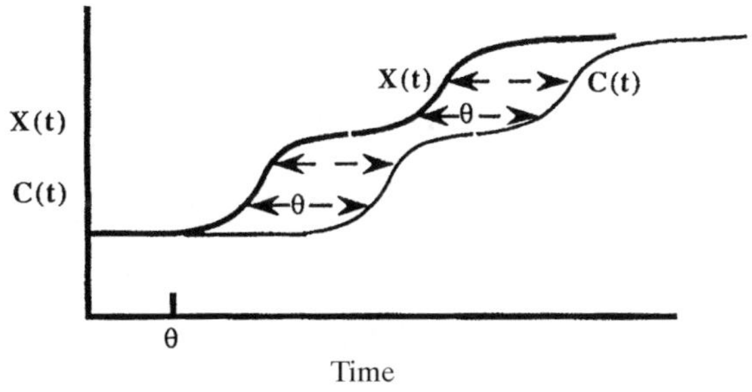

FIG. 8-18 Response of dead-time system.

Higher-Order Lags If a process is described by a series of n first-order lags, the overall system response becomes proportionally slower with each lag added. The special case of a series of n first-order lags with equal time constants has a transfer function given by

$$G(s) = \frac{K}{(\tau s + 1)^n} \tag{8-25}$$

The step response of this transfer function is shown in Fig. 8-19. Note that all curves reach about 60 percent of their final value at $t = n\tau$.

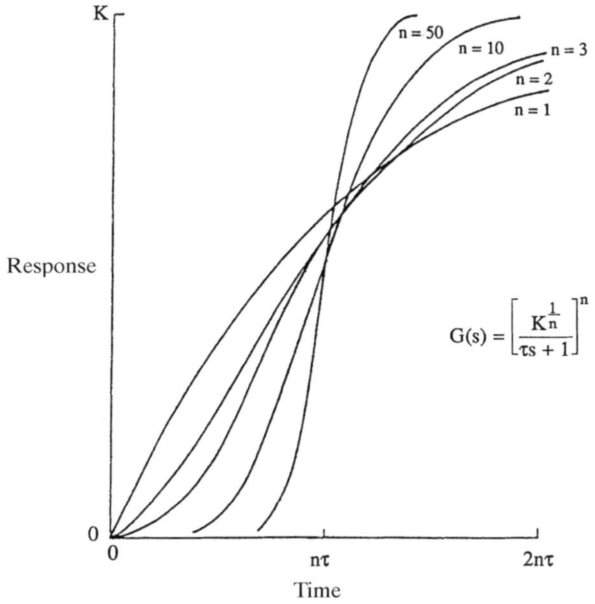

FIG. 8-19 Response of nth-order lags.

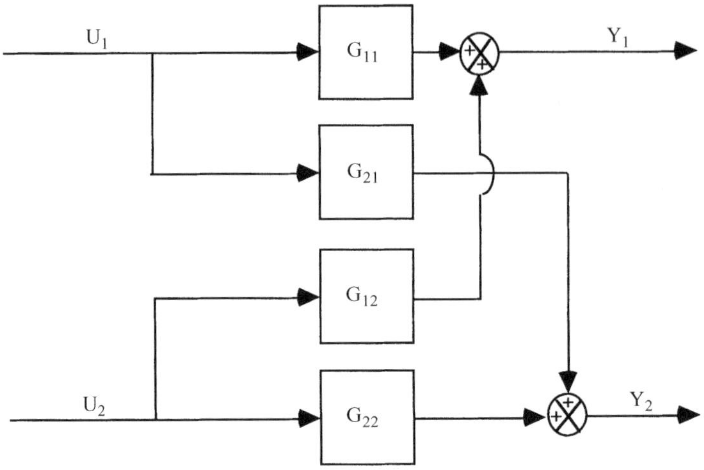

FIG. 8-20 Example of 2 × 2 transfer function.

Higher-order systems can be approximated by a first- or second-order plus dead-time system for control system design.

Multi-input, Multioutput Systems The dynamic systems considered up to this point have been examples of single-input, single-output (SISO) systems. In chemical processes, one often encounters systems where one input can affect more than one output. For example, assume that one is studying a distillation tower in which both reflux and boil-up are manipulated for control purposes. If the output variables are the top and bottom product compositions, then each input affects both outputs. For this distillation example, the process is referred to as a 2 × 2 system to indicate the number of inputs and outputs. In general, multi-input, multioutput (MIMO) systems can have n inputs and m outputs with $n \neq m$, and they can be nonlinear. Such a system would be called an $n \times m$ system. An example of a transfer function for a 2 × 2 linear system is given in Fig. 8-20. Note that since linear systems are involved, the effects of the two inputs on each output are additive. In many process control systems, one input is selected to control one output in a MIMO system. For m outputs there would be m such selections. For this type of control strategy, one needs to consider which inputs and outputs to couple, and this problem is referred to as loop pairing. Another important issue that arises involves interaction between control loops. When one loop makes a change in its manipulated variable, the change affects the other loops in the system. These changes are the direct result of the multivariable nature of the process. In some cases, the interaction can be so severe that overall control system performance is drastically reduced. Finally, some of the modern approaches to process control tackle the MIMO problem directly, and they simultaneously use all manipulated variables to control all output variables rather than pair one input to one output (see later section on multivariable control).

Fitting Dynamic Models to Experimental Data In developing empirical transfer functions, it is necessary to identify model parameters from experimental data. There are a number of approaches to process identification that have been published. The simplest approach involves introducing a step test into the process and recording the response of the process, as illustrated in Fig. 8-21. The x's in the figure represent the recorded data. For purposes of illustration, the process under study will be assumed to be first-order with dead time and have the transfer function

$$G(s) = \frac{Y(s)}{U(s)} = K \exp(-\theta s)/\tau s + 1 \qquad (8\text{-}26)$$

The response $y(t)$, produced by Eq. (8-26) can be found by inverting the transfer function, and it is also shown in Fig. 8-21 for a set of model parameters K, τ, and θ fitted to the data. These parameters are calculated by using optimization to minimize the squared difference between the model predictions and the data, i.e., a least squares approach. Let each measured data point be represented by y_j (measured response), t_j (time of measured response), $j = 1$ to n. Then the least squares problem can be formulated as

$$\min_{\tau,\theta,K} \sum_{j=1}^{n} [y_j - \hat{y}(t_j)]^2 \qquad (8\text{-}27)$$

where $\hat{y}(t_j)$ is the predicted value of y at time t_j and n is the number of data points. This optimization problem can be solved to calculate the optimal values of K, τ, and θ. A number of software packages such as Excel Solver are available for minimizing Eq. (8-27).

One operational problem caused by step forcing is the fact that the process under study is moved away from its steady-state operating point. Plant managers may be reluctant to allow large steady-state changes, since normal production will be disturbed by the changes. As a result, alternative methods of forcing actual processes have been developed, and these included pulse testing and pseudo-random binary signal (PRBS) forcing, both of which are illustrated in Fig. 8-22. With pulse forcing, one introduces a step, and then after a period of time the input is returned to its original value. The result is that the process dynamics are excited, but after the forcing the process returns to its original steady state. PRBS forcing involves a series of pulses of fixed height and random duration, as shown in Fig. 8-22. The advantage of PRBS is that forcing can be concentrated on particular frequency ranges that are important for control system design.

Transfer function models are linear, but chemical processes are known to exhibit nonlinear behavior. One could use the same type of optimization objective as given in Eq. (8-27) to determine parameters in nonlinear first-principles models, such as Eq. (8-3) presented earlier. Also, nonlinear empirical models, such as neural network models, have recently been proposed for process applications. The key to the use of these nonlinear empirical models is to have high-quality process data, which allows the important nonlinearities to be identified.

FEEDBACK CONTROL SYSTEM CHARACTERISTICS

GENERAL REFERENCES: Shinskey, *Process Control Systems*, 4th ed., McGraw-Hill, New York, 1996; Seborg, Edgar, and Mellichamp, *Process Dynamics and Control*, Wiley, New York, 1989.

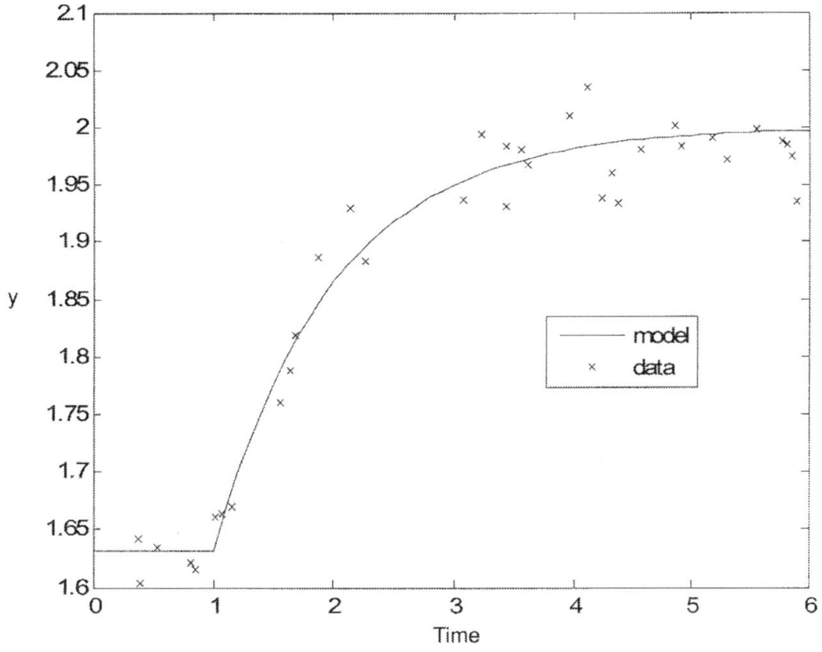

FIG. 8-21 Plot of experimental data and first-order model fit.

There are two objectives in applying feedback control: (1) regulate the controlled variable at the set point following changes in load and (2) respond to set-point changes, the latter called servo operation. In fluid processes, almost all control loops must contend with variations in load, and therefore regulation is of primary importance. While most loops will operate continuously at fixed set points, frequent changes in set points can occur in flow loops and in batch production. The most common mechanism for achieving both objectives is feedback control, because it is the simplest and most universally applicable approach to the problem.

Closing the Loop The simplest representation of the closed feedback loop is shown in Fig. 8-23. The load is shown entering the process at the same point as the manipulated variable because that is the most common point of entry, and because, lacking better information, the elements in the path of the manipulated variable are the best estimates of those in the load path. The load rarely impacts directly on the controlled variable without passing through the dominant lag in the process. Where the load is unmeasured, its current value can be observed as the controller output required to keep the controlled variable Y at set point Y_{sp}.

If the loop is opened—either by placing the controller in manual operation or by setting its gains to zero—the load will have complete influence over the controlled variable, and the set point will have none. Only by closing the loop with controller gain as high as possible will the influence of the load be minimized and that of the set point be maximized. There is a practical limit to the controller gain, however, at the point where the controlled variable develops a uniform oscillation. This is defined as the limit of stability, and it is reached when the product of gains in the loop $\Pi G = G_c G_v G_p$ is equal to 1.0 at the period of the oscillation. If the gain of any element in the loop increases from this condition, oscillations will expand, creating a dangerous situation where safe limits of operation could be exceeded in a few cycles. Consequently, control loops should be left in a condition where the loop gain is less than 1.0 by a safe margin that allows for possible variations in process parameters. Figure 8-24 describes a load response under PID (proportional-integral-derivative) control where the loop is well damped at a loop gain of 0.56; loop gain is then increased to 0.93 and to 1.05, creating a lightly damped and then an expanding cycle, respectively.

In controller tuning, a choice must be made between performance and robustness. Performance is a measure of how well a given controller with certain parameter settings regulates a variable, relative to the best response that can be achieved for that particular process. Robustness is a measure of how small a change in a process parameter is required to bring the loop from its current state to the limit of stability ($\Pi G = 1.0$). The well-damped loop in Fig. 8-24 has a robustness of 79 percent, in that increasing the gain of any element in the loop by a factor of 1/0.56, or 1.79, would bring the loop to the limit of stability. Increasing controller performance by raising its gain can therefore be expected to decrease robustness. Both performance and robustness are functions of the dynamics of the process being controlled, the selection of the controller, and the tuning of the controller parameters.

On/Off Control An on/off controller is used for manipulated variables having only two states. They commonly control temperatures in homes, electric water heaters and refrigerators, and pressure and liquid level in pumped storage systems. On/off control is satisfactory where slow cycling is acceptable, because it always leads to cycling when the load lies between the two states of the manipulated variable. The cycle will be positioned symmetrically about the set point only if the load happens to be equidistant between the two states of the manipulated variable. The period of the symmetric cycle will be approximately 4θ, where θ is the dead time in the loop. If the load is not centered between the states of the manipulated variable, the period will tend to increase and the cycle will follow a sawtooth pattern.

Every on/off controller has some degree of dead band, also known as lockup, or differential gap. Its function is to prevent erratic switching between states, thereby extending the life of contacts and motors. Instead of changing states precisely when the controlled variable crosses the set point, the controller will change states at two different points for increasing and decreasing signals. The difference between these two switching points is the dead band (see Fig. 8-25); it increases the amplitude and period of the cycle, similar to the effects of dead time.

A three-state controller is used to drive either a pair of independent two-state actuators, such as heating and cooling valves, or a bidirectional motorized actuator. The controller is comprised of two on/off controllers, each with dead band, separated by a dead zone. While the controlled variable lies within the dead zone, neither output is energized. This controller can drive a motorized valve to the point where the manipulated variable matches the load, thereby avoiding cycling.

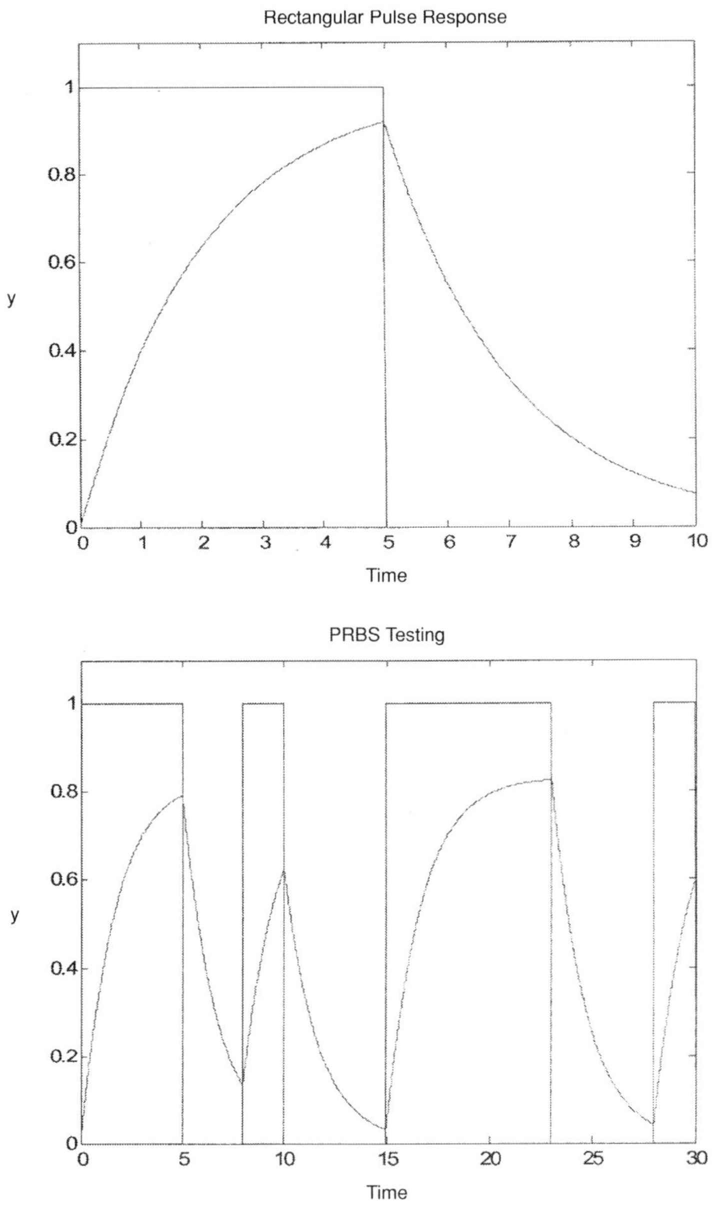

FIG. 8-22 Rectangular pulse response and PRBS testing.

Proportional Control A proportional controller moves its output proportional to the deviation e between the controlled variable y and its set point y_{sp}:

$$u = K_c e + b = \frac{100}{P} e + b \qquad (8\text{-}28)$$

where $e = \pm(y - y_{sp})$, the sign selected to produce negative feedback. In some controllers, proportional gain K_c is introduced as a pure number; in others, it is set as $100/P$, where P is the proportional band in percent. The output bias b of the controller is also known as manual reset. The proportional controller is not a good regulator, because any change in output required to respond to a change in load results in a corresponding change in the controlled variable. To minimize the resulting offset, the bias should be set at the best estimate of the load, and the proportional band set as low as possible. Processes requiring a proportional band of more than a few percent may control with unacceptably large values of offset.

Proportional control is most often used to regulate liquid level, where variations in the controlled variable carry no economic penalty and where other control modes can easily destabilize the loop. It is actually recommended for controlling the level in a surge tank when manipulating the flow of feed to a critical downstream process. By setting the proportional band just under 100 percent, the level is allowed to vary over the full range of the tank capacity as inflow fluctuates, thereby minimizing the resulting rate of change of manipulated outflow. This technique is called averaging level control.

Proportional-plus-Integral (PI) Control Integral action eliminates the offset described above by moving the controller output at a

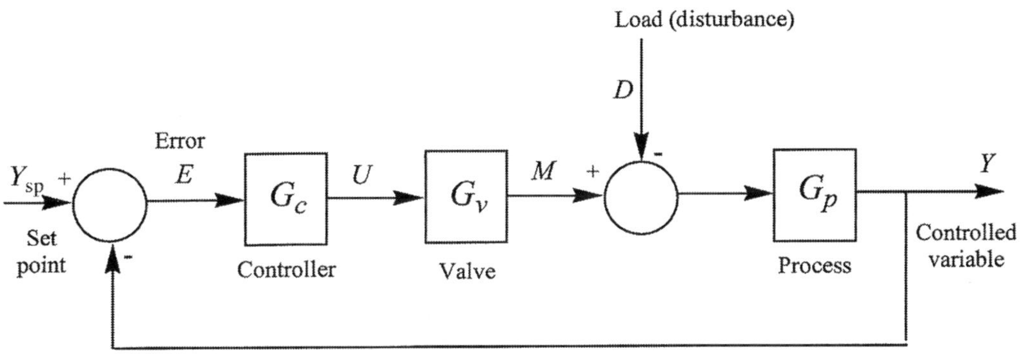

FIG. 8-23 Both load regulation and set-point response require high gains for the feedback controller.

rate proportional to the deviation from set point—the output will then not stop moving until the deviation is zero. Although available alone in an integral controller, it is most often combined with proportional action in a PI controller:

$$u = \frac{100}{P}\left(e + \frac{1}{\tau_I}\int e dt\right) + C_0 \qquad (8\text{-}29)$$

where τ_I is the integral time constant in minutes; in some controllers it is introduced as integral gain or reset rate $1/\tau_I$ in repeats per minute. The last term in the equation is the constant of integration, the value of the controller output when integration begins. The PI controller is by far the most commonly used controller in the process industries.

Because the integral term lags the proportional term by 90° in phase, the PI controller then always produces a phase lag between 0° and 90°:

$$\phi_{PI} = -\tan^{-1}\frac{\tau_o}{2\pi\tau_I} \qquad (8\text{-}30)$$

where τ_o is the period of oscillation of the loop. The phase angle should be kept between 15° for lag-dominant processes and 45° for dead-time-dominant processes for optimum results.

Proportional-plus-Integral-plus-Derivative (PID) Control The derivative mode moves the controller output as a function of the rate of change of the controlled variable, which adds phase lead to the controller, increasing its speed of response. It is normally combined with proportional and integral modes. The noninteracting or ideal form of the PID controller appears functionally as

$$u = \frac{100}{P}\left(e + \frac{1}{\tau_I}\int e\, dt \pm \tau_D\frac{dy}{dt}\right) + C_0 \qquad (8\text{-}31)$$

where τ_D is the derivative time constant. Note that derivative action is applied to the controlled variable rather than to the deviation, as it should not be applied to the set point; the selection of the sign for the derivative term must be consistent with the action of the controller.

In some PID controllers, the integral and derivative terms are combined serially rather than in parallel, as done in the last equation. This results in interaction between these modes, such that the effective values of the controller parameters differ from their set values as follows:

$$\tau_{I,\text{eff}} = \tau_I + \tau_D$$

$$\tau_{D,\text{eff}} = \frac{1}{1/\tau_D + 1/\tau_I} \qquad (8\text{-}32)$$

$$K_c = \frac{100}{P}\left(1 + \frac{\tau_D}{\tau_I}\right)$$

The performance of the interacting controller is almost as high as that of the noninteracting controller on most processes, but the tuning rules differ because of the above relationships. Both controllers are in common use in digital systems.

There is always a gain limit placed upon the derivative vector—a value of 10 is typical. However, interaction decreases the derivative

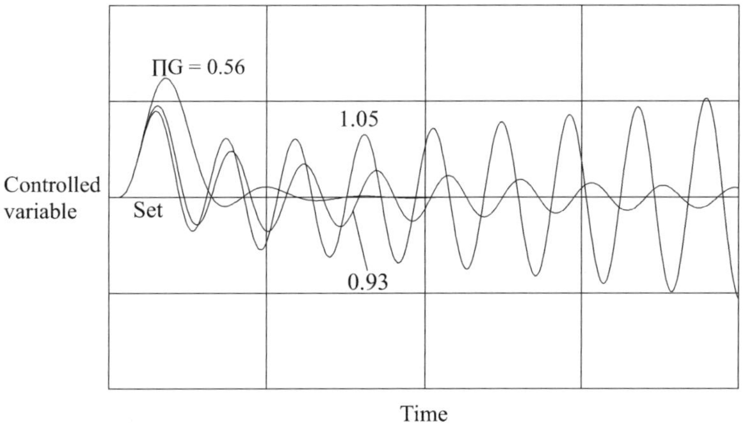

FIG. 8-24 Transition from well-damped load response to instability develops as loop gain increases.

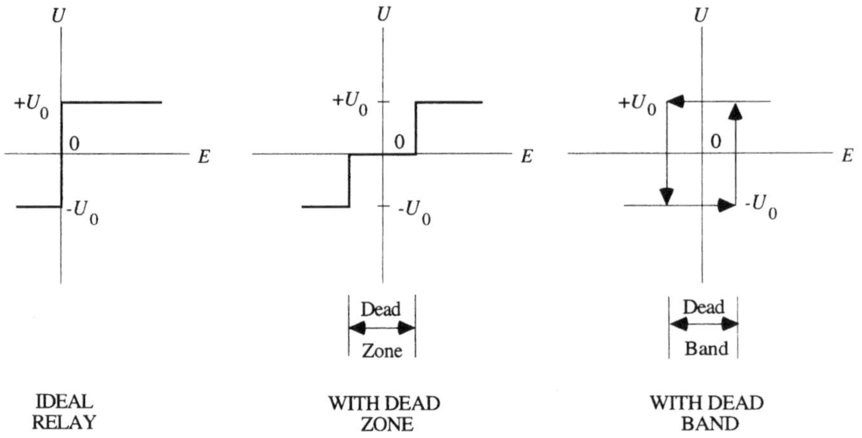

FIG. 8-25 On/off controller characteristics.

gain below this value by the factor $1 + \tau_D/\tau_I$, which is the reason for the decreased performance of the interacting PID controller. Sampling in a digital controller has a similar effect, limiting derivative gain to the ratio of derivative time to the sample interval of the controller. Noise on the controlled variable is amplified by the derivative gain, preventing its use in controlling flow and liquid level. Derivative action is recommended for control of temperature and composition in multiple-capacity processes with little measurement noise.

Controller Comparison Figure 8-26 compares the step load response of a distributed lag without control, and with P, PI, and interacting PID control. A distributed lag is a process whose resistance and capacity are distributed throughout its length—a heat exchanger is characteristic of this class, its heat-transfer surface and heat capacity being uniformly distributed. Other examples include imperfectly stirred tanks and distillation columns—both trayed and packed. The signature of a distributed lag is its open-loop (uncontrolled) step response, featuring a relatively short dead time followed by a dominant lag called $\Sigma\tau$, which is the time required to reach 63.2 percent complete response.

The proportional controller is unable to return the controlled variable to the set point following the step load change, as a deviation is required to sustain its output at a value different from its fixed bias b. The amount of proportional offset produced as a fraction of the uncontrolled offset is $1/(1 + KK_c)$, where K is the steady-state process gain—in Fig. 8-26, that fraction is 0.13. Increasing K_c can reduce the offset, but with an accompanying loss in damping.

The PI and PID controller were tuned to produce a minimum *integrated absolute error* (*IAE*). Their response curves are similar in appearance to a gaussian distribution curve, but with a damped cycle in the trailing edge. The peak deviation of the PID response curve is only 0.12 times the uncontrolled offset, occurring at $0.36\Sigma\tau$; the peak deviation of the PI response curve is 0.21 times the uncontrolled offset, occurring at $0.48\Sigma\tau$. These values can be used to predict the load response of any distributed lag whose parameters K and $\Sigma\tau$ are known or can be estimated as described below.

CONTROLLER TUNING

The performance of a controller depends as much on its tuning as on its design. Tuning must be applied by the end user to fit the controller to the controlled process. There are many different approaches to controller tuning, based on the particular performance criteria selected, whether load or set-point changes are more important, whether the process is lag- or dead-time-dominant, and the availability of information about the process dynamics. The earliest definitive work in this field was done at the Taylor Instrument Company by Ziegler and Nichols (*Trans. ASME*, p. 759, 1942), tuning PI and interacting PID

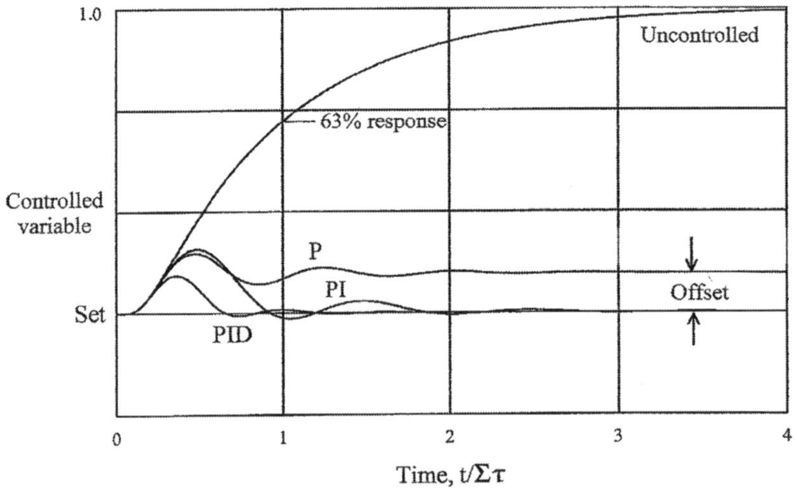

FIG. 8-26 Minimum-IAE tuning gives very satisfactory load response for a distributed lag.

controllers for optimum response to step load changes applied to lag-dominant processes. While these tuning rules are still in use, they do not apply to set-point changes, dead-time-dominant processes, or noninteracting PID controllers (Seborg, Edgar, and Mellichamp, *Process Dynamics and Control*, Wiley, New York, 2004).

Controller Performance Criteria The most useful measures of controller performance in an industrial setting are the maximum deviation in the controlled variable resulting from a disturbance, and its integral. The disturbance could be to the set point or to the load, depending on the variable being controlled and its context in the process. The size of the deviation and its integral are proportional to the size of the disturbance (if the loop is linear at the operating point). While actual disturbances arising in a plant may appear to be random, the controller needs a reproducible test to determine how well it is tuned. The disturbance of choice for test purposes is the step, because it can be applied manually, and by containing all frequencies including zero it exercises all modes of the controller. (The step actually has the same frequency distribution as integrated white noise, a "random walk.") When tuned optimally for step disturbances, the controller will be optimally tuned for most other disturbances as well.

A step change in set point, however, may be a poor indicator of a loop's load response. For example, a liquid-level controller does not have to integrate to follow a set-point change, as its steady-state output is independent of the set point. Stepping a flow controller's set point is an effective test of its tuning, however, as its steady-state output is proportional to its set point. Other loops should be load-tested: simulate a load change from a steady state at zero deviation by transferring the controller to manual and stepping its output, and then immediately transferring back to automatic before a deviation develops.

Figure 8-27*a* and *b* shows variations in the response of a distributed lag to a step change in load for different combinations of proportional and integral settings of a PI controller. The maximum deviation is the most important criterion for variables that could exceed safe operating levels, such as steam pressure, drum level, and steam temperature in a boiler. The same rule can apply to product quality if violating specifications causes it to be rejected. However, if the product can be accumulated in a downstream storage tank, its average quality is more important, and this is a function of the deviation integrated over the residence time of the tank. Deviation in the other direction, where the product is better than specification, is safe but increases production costs in proportion to the integrated deviation because quality is given away.

For a PI or PID controller, the integrated deviation—better known as integrated error IE—is related to the controller settings

$$\mathrm{IE} = \Delta u \frac{P\tau_I}{100} \tag{8-33}$$

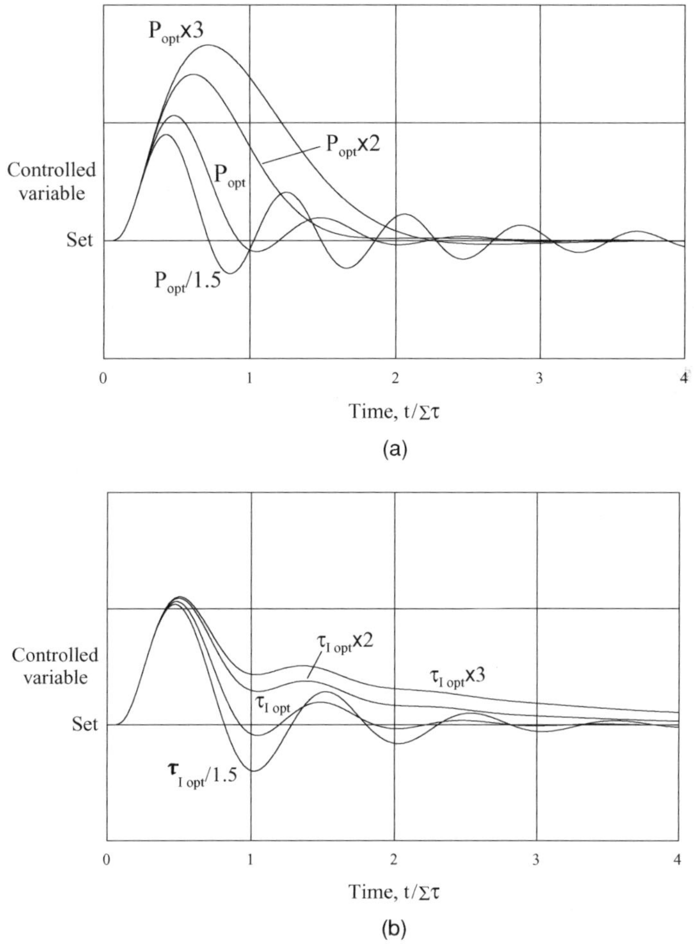

FIG. 8-27 The optimum settings produce minimum-IAE load response. (*a*) The proportional band primarily affects damping and peak deviation. (*b*) Integral time determines overshoot.

where Δu is the difference in controller outputs between two steady states, as required by a change in load or set point. The proportional band P and integral time τ_I are the indicated settings of the controller for PI and both interacting and noninteracting PID controllers. Although the derivative term does not appear in the relationship, its use typically allows a 50 percent reduction in integral time and therefore in IE. The integral time in the IE expression should be augmented by the sample interval if the controller is digital, the time constant of any filter used, and the value of any dead-time compensator.

It would appear, from the above, that minimizing IE is simply a matter of minimizing the P and τ_I settings of the controller. However, settings will be reached that produce excessive oscillations, such as shown in the lowest two response curves in Fig. 8-27*a* and *b*. It is preferable instead to find a combination of controller settings that minimizes integrated absolute error IAE, which for both load and set-point changes is a well-damped response with minimal overshoot. The curves designated P_{opt} and $\tau_{I,opt}$ in Fig. 8-27 are the same minimum-IAE response to a step change in load for a distributed-lag process under PI control. Because of the very small overshoot, the IAE will be only slightly larger than the IE. Loops that are tuned to minimize IAE tend to give responses that are close to minimum IE and with minimum peak deviation. The other curves in Fig. 8-27*a* and *b* describe the effects of individual adjustments to P and τ_I, respectively, around those optimum values and can serve as a guide to fine-tuning a PI controller.

The performance of a controller (and its tuning) must be based on what is achievable for a given process. The concept of best practical IE (IE_b) for a step change in load ΔL to a process consisting of dead time and one or two lags can be estimated (Shinskey, *Process Control Systems*, 4th ed., McGraw-Hill, New York, 1996)

$$IE_b = \Delta L K_L \tau_L (1 - e^{-\theta/\tau_L}) \tag{8-34}$$

where K_L is the gain and τ_L the primary time constant in the load path, and θ the dead time in the manipulated path to the controlled variable. If the load or its gain is unknown, Δu and K ($= K_v K_p$) may be substituted. If the process is non-self-regulating (i.e., an integrator), the relationship is

$$IE_b = \frac{\Delta L \theta^2}{\tau_1} \tag{8-35}$$

where τ_1 is the time constant of the process integrator. The peak deviation of the best practical response curve is

$$e_b = \frac{IE_b}{\theta + \tau_2} \tag{8-36}$$

where τ_2 is the time constant of a common secondary lag (e.g., in the measuring device).

TABLE 8-2 Tuning Rules Using Known Process Parameters

Process	Controller	P	τ_I	τ_D
Dead-time-dominant	PI	$250K$	0.5θ	
Lag-dominant	PI	$106K\theta/\tau_m$	4.0θ	
	PID_n	$77K\theta/\tau_m$	1.8θ	0.45θ
	PID_i	$106K\theta/\tau_m$	1.5θ	0.55θ
Non-self-regulating	PI	$106\theta/\tau_1$	4.0θ	
	PID_n	$78\theta/\tau_1$	1.9θ	0.48θ
	PID_i	$108\theta/\tau_1$	1.6θ	0.58θ
Distributed lags	PI	$20K$	$0.50\ \Sigma\tau$	
	PID_n	$10K$	$0.30\ \Sigma\tau$	$0.09\ \Sigma\tau$
	PID_i	$15K$	$0.25\ \Sigma\tau$	$0.10\Sigma\tau$

NOTE: n = noninteracting, i = interacting controller modes.

The performance of any controller can be measured against this standard by comparing the IE it achieves in responding to a load change with the best practical IE. Maximum performance levels for PI controllers on lag-dominant processes lie in the 20 to 30 percent range, while for PID controllers they fall between 40 and 60 percent, varying with secondary lags.

Tuning Methods Based on Known Process Models The most accurate tuning rules for controllers have been based on simulation, where the process parameters can be specified and IAE and IE can be integrated during the simulation as an indication of performance. Controller settings are then iterated until a minimum IAE is reached for a given disturbance. Next these optimum settings are related to the parameters of the simulated process in tables, graphs, or equations, as a guide to tuning controllers for processes whose parameters are known (Seborg, Edgar, and Mellichamp, *Process Dynamics and* Control, Wiley, New York, 2004). This is a multidimensional problem, however, in that the relationships change as a function of process type, controller type, and source of disturbance.

Table 8-2 summarizes these rules for minimum-IAE load response for the most common controllers. The process gain and time constant τ_m are obtained from the product of G_v and G_p in Fig. 8-23. Derivative action is not effective for dead-time-dominant processes. Any secondary lag, sampling interval, or filter time constant should be added to dead time θ.

The principal limitation to using these rules is that the true process parameters are often unknown. Steady-state gain K can be calculated from a process model, or determined from the steady-state results of a step test as $\Delta c/\Delta u$, as shown in Fig. 8-28. The test will not be viable, however if the time constant of the process τ_m is longer than a few minutes, since five time constants must elapse to approach a steady state within 1 percent, and unexpected disturbances may intervene. Estimated dead time θ is the time from the step to the intercept of a

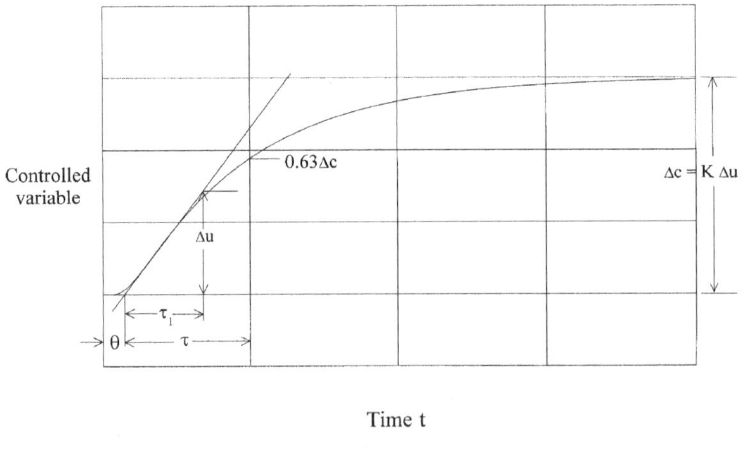

FIG. 8-28 If a steady state can be reached, gain K and time constant τ can be estimated from a step response; if not, use τ_1 instead.

TABLE 8-3 Tuning Rules Using Slope and Intercept

Controller	P	τ_I	τ_D
PI	$150\theta/\tau_1$	3.5θ	
PID$_n$	$75\theta/\tau_1$	2.1θ	0.63θ
PID$_i$	$113\theta/\tau_1$	1.8θ	0.70θ

NOTE: n = noninteracting, i = interacting controller modes.

TABLE 8-4 Tuning Rules Using Proportional Cycle

Controller	P	τ_I	τ_D
PI	$1.70P_u$	$0.81\tau_n$	
PID$_n$	$1.30P_u$	$0.48\tau_n$	$0.11\tau_n$
PID$_i$	$1.80P_u$	$0.38\tau_n$	$0.14\tau_n$

NOTE: n = noninteracting, i = interacting controller modes.

straight line tangent to the steepest part of the response curve. The estimated time constant τ is the time from that point to 63 percent of the complete response. In the presence of a significant secondary lag, these results will not be completely accurate, however. The time for 63 percent response may be more accurately calculated as the residence time of the process: its volume divided by current volumetric flow rate.

Tuning Methods When Process Model Is Unknown Ziegler and Nichols developed two tuning methods for processes with unknown parameters. The open-loop method uses a step test without waiting for a steady state to be reached and is therefore applicable to very slow processes. Dead time is estimated from the intercept of the steepest tangent to the response curve in Fig. 8-28, whose slope is also used. If the process is non-self-regulating, the controlled variable will continue to follow this slope, changing by an amount equal to Δu in a time equal to its time constant τ_1. This time estimate τ_1 is used along with θ to tune controllers according to Table 8-3, applicable to lag-dominant processes. If the process is known to be a distributed lag, such as a heat exchanger, distillation column, or stirred tank, then better results will be obtained by converting the estimated values of θ and τ_1 to K and $\Sigma\tau$ and using Table 8-2. The conversion factors are $K = 7.5\theta/\tau_1$ and $\Sigma\tau = 7.0\theta$.

The Ziegler and Nichols closed-loop method requires forcing the loop to cycle uniformly under proportional control, by setting the integral time to maximum and derivative time to zero and reducing the proportional band until a constant-amplitude cycle results. The natural period τ_n of the cycle (the proportional controller contributes no phase shift to alter it) is used to set the optimum integral and derivative time constants. The optimum proportional band is set relative to the undamped proportional band P_u which was found to produce the uniform oscillation. Table 8-4 lists the tuning rules for a lag-dominant process.

A uniform cycle can also be forced by using on-off control to cycle the manipulated variable between two limits. The period of the cycle will be close to τ_n if the cycle is symmetric; the peak-to-peak amplitude A_c of the controlled variable divided by the difference between the output limits A_u is a measure of process gain at that period and is therefore related to P_u for the proportional cycle:

$$P_u = 100\frac{\pi}{4}\frac{A_c}{A_u} \tag{8-37}$$

The factor $\pi/4$ compensates for the square wave in the output. Tuning rules are given in Table 8-4.

Set-Point Response All the above tuning methods are intended to minimize IAE for step load changes. When applied to lag-dominant processes, the resulting controller settings produce excessive overshoot of set-point changes. This behavior has led to the practice of tuning to optimize set-point response, which unfortunately degrades the load response of lag-dominant loops. An option has been available with some controllers to remove proportional action from set-point changes, which eliminates set-point overshoot but lengthens settling time. A preferred solution to this dilemma is available in many modern controllers which feature an independent gain adjustment for the set point, through which set-point response can be optimized after the controller has been tuned to optimize load response.

Figure 8-29 shows set-point and load responses of a distributed lag for both set-point and load tuning, including the effects of fractional set-point gain K_r. The set point was stepped at time zero, and the load stepped at time 2.4. With full set-point gain, the PI controller was tuned for minimum-IAE set-point response with $P = 29K$ and $\tau_I = \Sigma\tau$, compared to $P = 20K$ and $\tau_I = 0.50\Sigma\tau$ for minimum-IAE load response. These settings increase its IE for load response by a factor of 2.9, and its peak deviation by 20 percent, over optimum load tuning. However, with optimum load tuning, that same set-point overshoot can be obtained with set-point gain $K_r = 0.54$. The effects of full set-point gain (1.0) and no set-point gain (0) are shown for comparison.

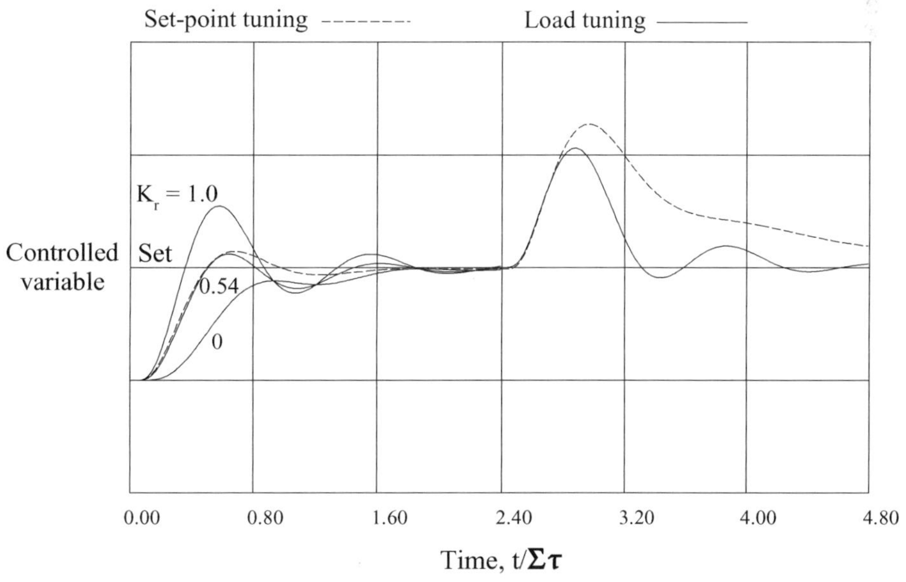

FIG. 8-29 Tuning proportional and integral settings to optimize set-point response degrades load response; using a separate set-point gain adjustment allows both responses to be optimized.

ADVANCED CONTROL SYSTEMS

BENEFITS OF ADVANCED CONTROL

The economics of most processes are determined by the steady-state operating conditions. Excursions from these steady-state conditions usually have a less important effect on the economics of the process, except when the excursions lead to off-specification products. To enhance the economic performance of a process, the steady-state operating conditions must be altered in a manner that leads to more efficient process operation.

The hierarchy shown in Fig. 8-30 indicates that process control activities consist of the following five levels:

Level 1: Measurement devices and actuators
Level 2: Safety, environmental/equipment protection
Level 3: Regulatory control
Level 4: Real-time optimization
Level 5: Planning and scheduling

Levels 4 and 5 clearly affect the process economics, as both levels are directed to optimizing the process in some manner. In contrast, levels 1, 2, and 3 would appear to have no effect on process economics. Their direct effect is indeed minimal, although indirectly they can have a major effect. Basically, these levels provide the foundation for all higher levels. A process cannot be optimized until it can be operated consistently at the prescribed targets. Thus, satisfactory regulatory control must be the first goal of any automation effort. In turn, the measurements and actuators provide the process interface for regulatory control.

For most processes, the optimum operating point is determined by a constraint. The constraint might be a product specification (a product stream can contain no more than 2 percent ethane); violation of this constraint causes off-specification product. The constraint might be an equipment limit (vessel pressure rating is 300 psig); violation of this constraint causes the equipment protection mechanism (pressure

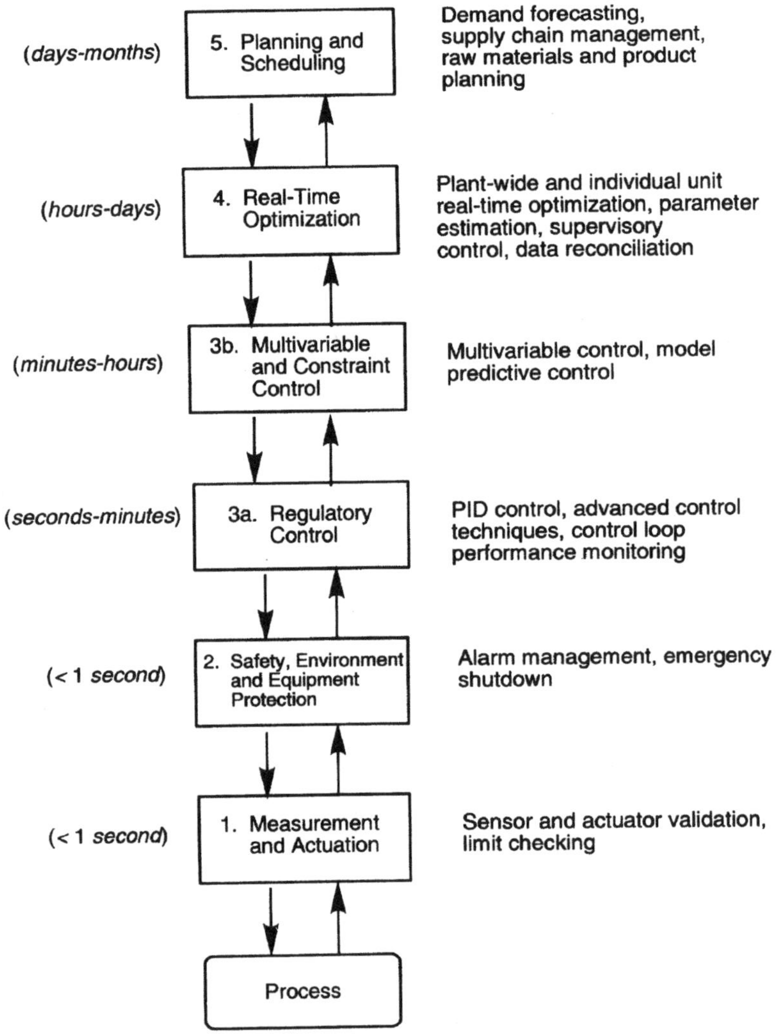

FIG. 8-30 The five levels of process control and optimization in manufacturing. Time scales are shown for each level. (*Source: Seborg et al.,* Process Dynamics and Control, *2d ed., Wiley, New York, 2004.*)

relief device) to activate. As the penalties are serious, violation of such constraints must be very infrequent.

If the regulatory control system were perfect, the target could be set exactly equal to the constraint (i.e., the target for the pressure controller could be set at the vessel relief pressure). However, no regulatory control system is perfect. Therefore, the value specified for the target must be on the safe side of the constraint, thus allowing the control system some operating margin. How much depends on the following:

1. *The performance of the control system (i.e., how effectively it responds to disturbances).* The faster the control system reacts to a disturbance, the closer the process can be operated to the constraint.
2. *The magnitude of the disturbances to which the control system must respond.* If the magnitude of the major disturbances can be reduced, the process can be operated closer to the constraint.

One measure of the performance of a control system is the variance of the controlled variable from the target. Both improving the control system and reducing the disturbances will lead to a lower variance in the controlled variable.

In a few applications, improving the control system leads to a reduction in off-specification product and thus improved process economics. However, in most situations, the process is operated sufficiently far from the constraint that very little, if any, off-specification product results from control system deficiencies. Management often places considerable emphasis on avoiding off-specification production, so consequently the target is actually set far more conservatively than it should be.

In most applications, simply improving the control system does not directly lead to improved process economics. Instead, the control system improvement must be accompanied by shifting the target closer to the constraint. There is always a cost of operating a process in a conservative manner. The cost may be a lower production rate, a lower process efficiency, a product giveaway, or other. When management places undue emphasis on avoiding off-specification production, the natural reaction is to operate very conservatively, thus incurring other costs.

The immediate objective of an advanced control effort is to reduce the variance in an important controlled variable. However, this effort must be coupled with a commitment to adjust the target for this controlled variable so that the process is operated closer to the constraint. In large-throughput (commodity) processes, very small shifts in operating targets can lead to large economic returns.

ADVANCED CONTROL TECHNIQUES

GENERAL REFERENCES: Seborg, Edgar, and Mellichamp, *Process Dynamics and Control,* Wiley, New York. 2004. Stephanopoulos, *Chemical Process Control: An Introduction to Theory and Practice,* Prentice-Hall, Englewood Cliffs, N.J., 1984. Shinskey, *Process Control Systems,* 4th ed., McGraw-Hill, New York, 1996. Ogunnaike and Ray, *Process Dynamics, Modeling, and Control,* Oxford University Press, New York, 1994.

While the single-loop PID controller is satisfactory in many process applications, it does not perform well for processes with slow dynamics, time delays, frequent disturbances, or multivariable interactions. We discuss several advanced control methods below that can be implemented via computer control, namely, feedforward control, cascade control, time-delay compensation, selective and override control, adaptive control, fuzzy logic control, and statistical process control.

Feedforward Control If the process exhibits slow dynamic response and disturbances are frequent, then the application of feedforward control may be advantageous. Feedforward (FF) control differs from feedback (FB) control in that the primary disturbance or load (D) is measured via a sensor and the manipulated variable (U) is adjusted so that deviations in the controlled variable from the set point are minimized or eliminated (see Fig. 8-31). By taking control action based on measured disturbances rather than controlled variable error, the controller can reject disturbances before they affect the controlled variable Y. To determine the appropriate settings for the manipulated variable, one must develop mathematical models that relate

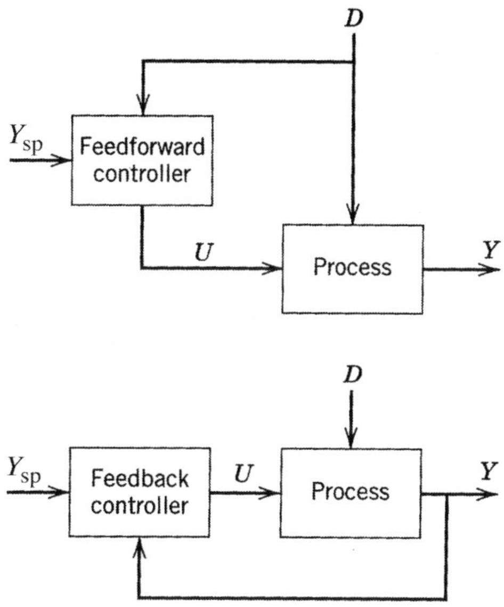

FIG. 8-31 Simplified block diagrams for feedforward and feedback control.

1. The effect of the manipulated variable U on the controlled variable Y
2. The effect of the disturbance D on the controlled variable Y

These models can be based on steady-state or dynamic analysis. The performance of the feedforward controller depends on the accuracy of both models. If the models are exact, then feedforward control offers the potential of perfect control (i.e., holding the controlled variable precisely at the set point at all times because of the ability to predict the appropriate control action). However, since most mathematical models are only approximate and since not all disturbances are measurable, it is standard practice to utilize feedforward control in conjunction with feedback control. Table 8-5 lists the relative advantages and disadvantages of feedforward and feedback control. By combining the two control methods, the strengths of both schemes can be utilized.

FF control therefore attempts to eliminate the effects of measurable disturbances, while FB control would correct for unmeasurable disturbances and modeling errors. This latter case is often referred to as feedback trim. These controllers have become widely accepted in the chemical process industries since the 1960s.

Design Based on Material and Energy Balances Consider a heat exchanger example (see Fig. 8-32) to illustrate the use of FF and FB control. The control objective is to maintain T_2, the exit liquid temperature, at the desired value (or set point) T_{2sp} despite variations in the inlet liquid flow rate F and inlet liquid temperature T_l. This is done by manipulating W, the steam flow rate. A feedback control scheme would entail measuring T_2, comparing T_2 to T_{2sp}, and then adjusting W. A feedforward control scheme requires measuring F and T_l, and adjusting W (knowing T_{2sp}), in order to control exit temperature T_2.

Figure 8-33*a* and *b* shows the control system diagrams for FB and FF control. A feedforward control algorithm can be designed for the heat exchanger in the following manner. Using a steady-state energy balance and assuming no heat loss from the heat exchanger,

$$WH = FC(T_2 - T_1) \tag{8-38}$$

where H = latent heat of vaporization and C = specific heat of liquid

$$W = \frac{C}{H} F(T_2 - T_1) \tag{8-39}$$

or

$$W = K_1 F(T_2 - T_1) \tag{8-40}$$

TABLE 8-5 Relative Advantages and Disadvantages of Feedforward and Feedback

Advantages	Disadvantages
Feedforward	
• Acts before the effect of a disturbance has been felt by the system • Is good for systems with large time constant or dead time • Does not introduce instability in the closed-loop response	• Requires direct measurement of all possible disturbances • Cannot cope with unmeasured disturbances • Is sensitive to process/model error
Feedback	
• Does not require identification and measurement of any disturbance for corrective action • Does not require an explicit process model • Is possible to design controller to be robust to process/model errors	• Control action not taken until the effect of the disturbance has been felt by the system • Is unsatisfactory for processes with large time constants and frequent disturbances • May cause instability in the closed-loop response

with

$$K_1 = \frac{C_L}{H} \qquad (8\text{-}41)$$

Replace T_2 by T_{2sp}

$$W = K_1 F(T_{set} - T_1) \qquad (8\text{-}42)$$

Equation (8-42) can be used in the FF calculation, assuming one knows the physical properties C and H. Of course, it is probable that the model will contain errors (e.g., unmeasured heat losses, incorrect C or H). Therefore, K_1 can be designated as an adjustable parameter that can be tuned. The use of a physical model for FF control is desirable because it provides a physical basis for the control law and gives an a priori estimate of what the tuning parameters should be. Note that such a model could be nonlinear [e.g., in Eq. (8-42), F and T_{2sp} are multiplied].

Block Diagram Analysis One shortcoming of this feedforward design procedure is that it is based on the steady-state characteristics of the process and, as such, neglects process dynamics (i.e., how fast the controlled variable responds to changes in the load and manipulated variables). Thus, it is often necessary to include "dynamic compensation" in the feedforward controller. The most direct method of designing the FF dynamic compensator is to use a block diagram of a general process, as shown in Fig. 8-34, where G_t represents the disturbance transmitter, G_f is the feedforward controller, G_d relates the disturbance to the controlled variable, G_v is the valve, G_p is the process, G_m is the output transmitter, and G_c is the feedback controller. All blocks correspond to transfer functions (via Laplace transforms).

Using block diagram algebra and Laplace transform variables, the controlled variable $Y(s)$ is given by

$$Y(s) = \frac{G_t G_f L(s) + G_d D(s)}{1 + G_m G_c G_v G_p} \qquad (8\text{-}43)$$

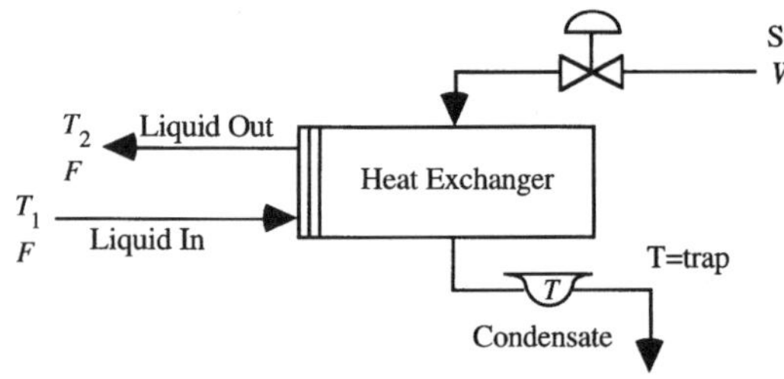

FIG. 8-32 A heat exchanger diagram.

For disturbance rejection $[D(s) \neq 0]$ we require that $Y(s) = 0$, or zero error. Solving Eq. (8-43) for G_f gives

$$G_f = \frac{-G_L}{G_t G_v G_p} \qquad (8\text{-}44)$$

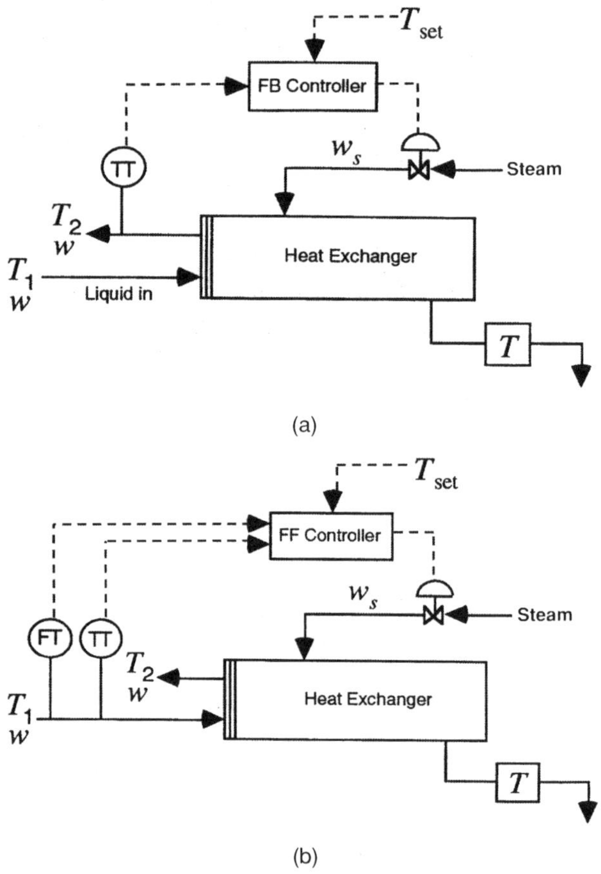

FIG. 8-33 (*a*) Feedback control of a heat exchanger. (*b*) Feedforward control of a heat exchanger.

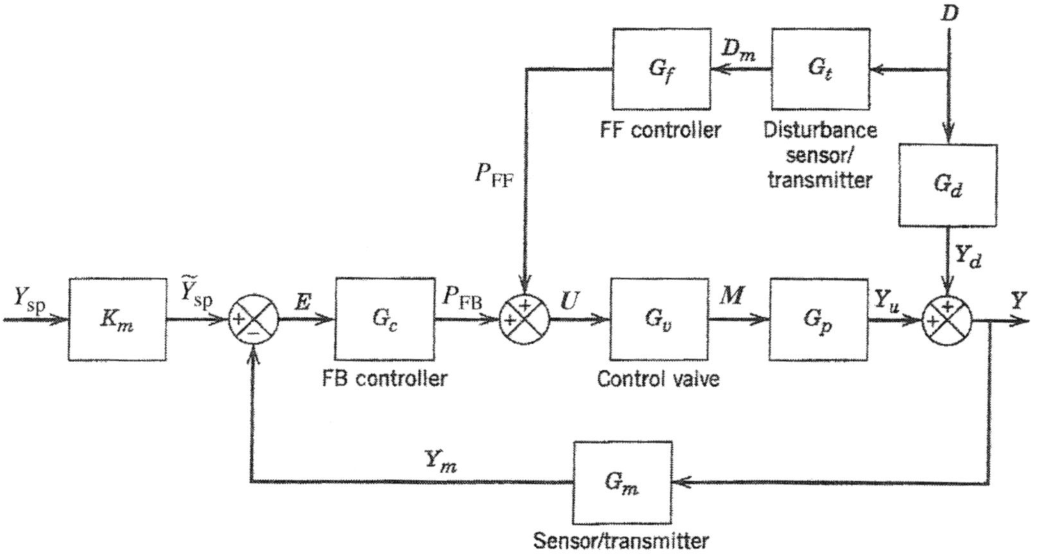

FIG. 8-34 A block diagram of a feedforward-feedback control system. (*Source: Seborg et al.,* Process Dynamics and Control, *2d ed., Wiley, New York, 2004.*)

Suppose the dynamics of G_d and G_p are first-order; in addition, assume that $G_c = K_c$ and $G_t = K_t$ (constant gains for simplicity).

$$G_d(s) = \frac{K_d}{\tau_d s + 1} = \frac{Y(s)}{D(s)} \tag{8-45}$$

$$G_p(s) = \frac{K_p}{\tau_p s + 1} = \frac{Y(s)}{U(s)} \tag{8-46}$$

Using Eq. (8-44),

$$G_f(s) = \frac{\tau_p s + 1}{\tau_d s + 1} \cdot \frac{-K_d}{K_p K_c K_t} = \frac{-K(\tau_p s + 1)}{\tau_d s + 1} \tag{8-47}$$

where K is the overall ratio of the gains in Eq. (8-47).

The above FF controller can be implemented by using a digital computer. Figure 8-35*a* and *b* compares typical responses for PID FB control, steady-state FF control ($s = 0$), dynamic FF control, and combined FF/FB control. In practice, the engineer can tune K, τ_p, and τ_d in the field to improve the performance of the FF controller. The feedforward controller can also be simplified to provide steady-state feedforward control. This is done by setting $s = 0$ in $G_f(s)$. This might be appropriate if there is uncertainty in the dynamic models for G_d and G_p.

Other Considerations in Feedforward Control The tuning of feedforward and feedback control systems can be performed independently. In analyzing the block diagram in Fig. 8-34, note that G_f is chosen to cancel out the effects of the disturbance $D(s)$, as long as there are no model errors. For the feedback loop, therefore, the effects of $D(s)$ can also be ignored, which for the servo case is

$$\frac{Y(s)}{Y_{sp}(s)} = \frac{G_c G_v G_p K_m}{1 + G_c G_v G_p G_m} \tag{8-48}$$

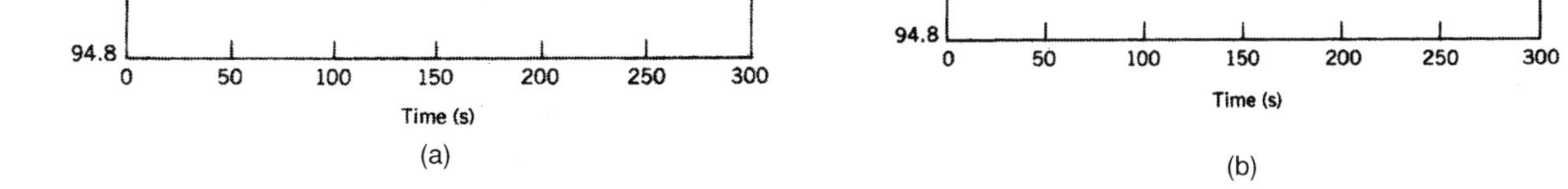

FIG. 8-35 (*a*) Comparison of FF (steady-state model) and PID FB control for disturbance change. (*b*) Comparison of FF (dynamic model) and combined FF/FB control.

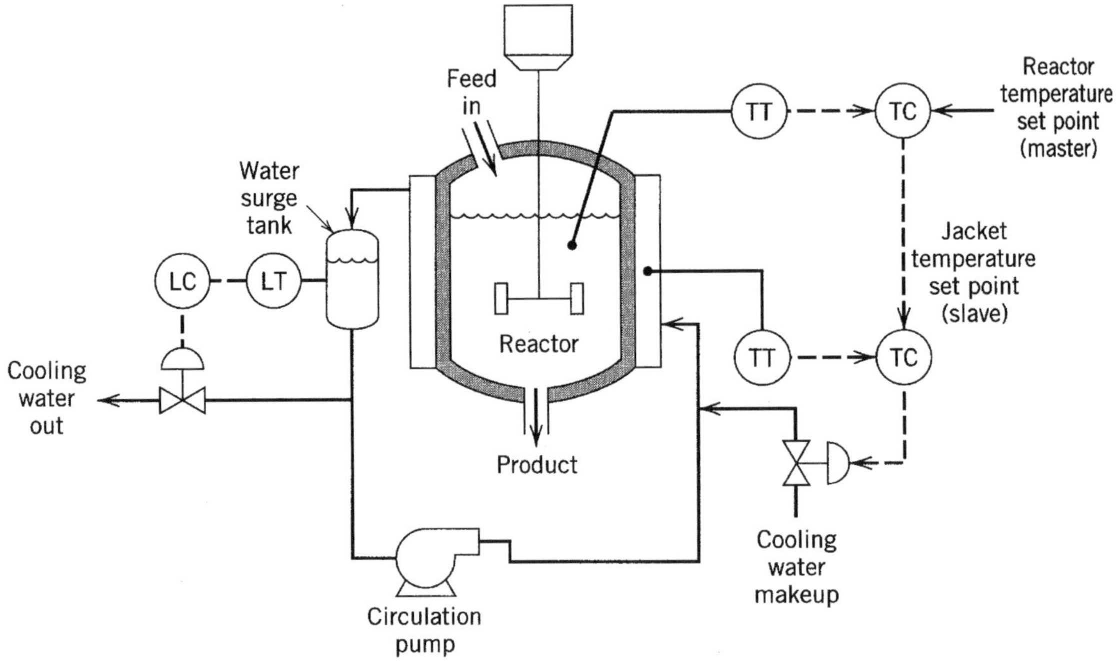

FIG. 8-36 Cascade control of an exothermic chemical reactor. (*Source: Seborg et al.,* Process Dynamics and Control, *2d ed., Wiley, New York, 2004.*)

Note that the characteristic equation will be unchanged for the FF + FB system; hence system stability will be unaffected by the presence of the FF controller. In general, the tuning of the FF controller can be less conservative than for the case of FB alone, because smaller excursions from the set point will result. This in turn would make the dynamic model $Y(s)$ more accurate.

The tuning of the controller in the feedback loop can be theoretically performed independently of the feedforward loop (i.e., the feedforward loop does not introduce instability in the closed-loop response). For more information on feedforward/feedback control applications and design of such controllers, refer to the general references.

Cascade Control One of the disadvantages of using conventional feedback control for processes with large time lags or delays is that disturbances are not recognized until after the controlled variable deviates from its set point. In these processes, correction by feedback control is generally slow and results in long-term deviation from the set point. One way to improve the dynamic response to load changes is by using a secondary measurement point and a secondary controller; the secondary measurement point is located so that it recognizes the upset condition before the primary controlled variable is affected.

One such approach is called cascade control, which is routinely used in most modern computer control systems. Consider a chemical reactor, where reactor temperature is to be controlled by coolant flow to the jacket of the reactor. The reactor temperature can be influenced by changes in disturbance variables such as feed rate or feed temperature; a feedback controller could be employed to compensate for such disturbances by adjusting a valve on the coolant flow to the reactor jacket. However, suppose an increase occurs in the coolant temperature as a result of changes in the plant coolant system. This will cause a change in the reactor temperature measurement, although such a change will not occur quickly, and the corrective action taken by the controller will be delayed.

Cascade control is one solution to this problem (see Fig. 8-36). Here the jacket temperature is measured, and an error signal is sent from this point to the coolant control valve; this reduces coolant flow, maintaining the heat-transfer rate to the reactor at a constant level and rejecting the disturbance. The cascade control configuration will also adjust the setting of the coolant control valve when an error occurs in the reactor temperature. The cascade control scheme shown in Fig. 8-36 contains two controllers. The primary controller is the reactor temperature controller. It measures the reactor temperature, compares it to the set point, and computes an output, which is the set point for the coolant flow rate controller. The secondary controller compares this set point to the coolant temperature measurement and adjusts the valve. The principal advantage of cascade control is that the secondary measurement (jacket temperature) is located closer to a potential disturbance in order to improve the closed-loop response.

Figure 8-37 shows the block diagram for a general cascade control system. In tuning of a cascade control system, the secondary controller (in the inner loop) is tuned first with the primary controller in manual. Often only a proportional controller is needed for the secondary loop, because offset in the secondary loop can be treated by using proportional-plus-integral action in the primary loop. When the primary controller is transferred to automatic, it can be tuned by using the techniques described earlier in this section. For more information on theoretical analysis of cascade control systems, see the general references for a discussion of applications of cascade control.

Time-Delay Compensation Time delays are a common occurrence in the process industries because of the presence of recycle loops, fluid-flow distance lags, and dead time in composition measurements resulting from use of chromatographic analysis. The presence of a time delay in a process severely limits the performance of a conventional PID control system, reducing the stability margin of the closed-loop control system. Consequently, the controller gain must be reduced below that which could be used for a process without delay. Thus, the response of the closed-loop system will be sluggish compared to that of the system with no time delay.

To improve the performance of time-delay systems, special control algorithms have been developed to provide time-delay compensation. The Smith predictor technique is the best-known algorithm; a related method is called the analytical predictor. Various investigators have

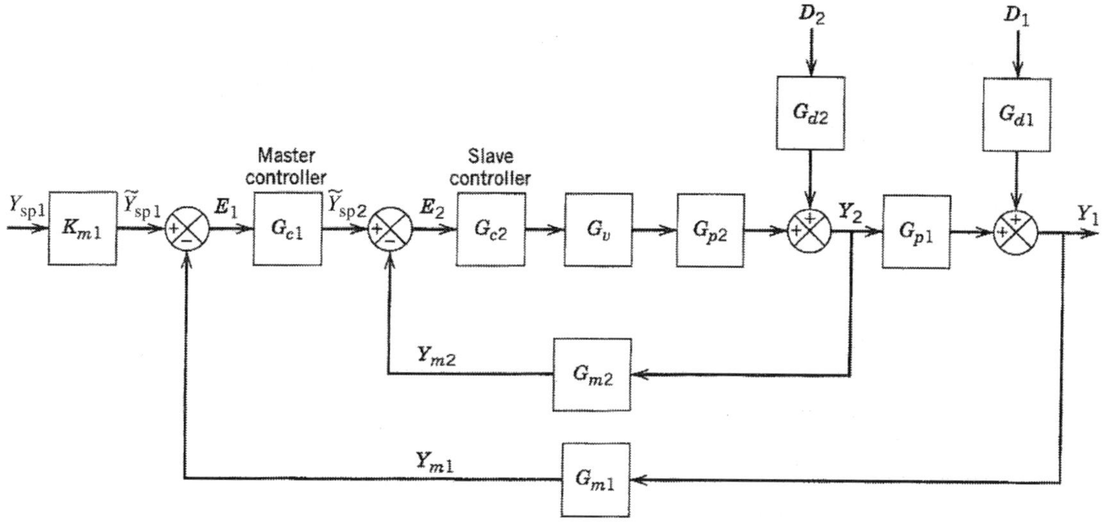

FIG. 8-37 Block diagram of the cascade control system. For a chemical reactor G_{d1} would correspond to a feed temperature or composition disturbance, while G_{d2} would be a change in the cooling water temperature. (*Source: Seborg et al.*, Process Dynamics and Control, *2d ed., Wiley, New York, 2004.*)

found that, based on integral squared error, the performance of the Smith predictor can be better than that for a conventional controller, as long as the time delay is known accurately.

The Smith predictor is a model-based control strategy that involves a more complicated block diagram than that for a conventional feedback controller, although a PID controller is still central to the control strategy (see Fig. 8-38). The key concept is based on better coordination of the timing of manipulated variable action. The loop configuration takes into account the fact that the current controlled variable measurement is not a result of the current manipulated variable action, but the value taken θ time units earlier. Time-delay compensation can yield excellent performance; however, if the process model parameters change (especially the time delay), the Smith predictor performance will deteriorate and is not recommended unless other precautions are taken.

Selective and Override Control When there are more controlled variables than manipulated variables, a common solution to this problem is to use a selector to choose the appropriate process variable from among a number of available measurements. Selectors can be based on multiple measurement points, multiple final control elements, or multiple controllers, as discussed below. Selectors are used to improve the control system performance as well as to protect equipment from unsafe operating conditions.

One type of selector device chooses as its output signal the highest (or lowest) of two or more input signals. This approach is often referred to as auctioneering. On instrumentation diagrams, the symbol

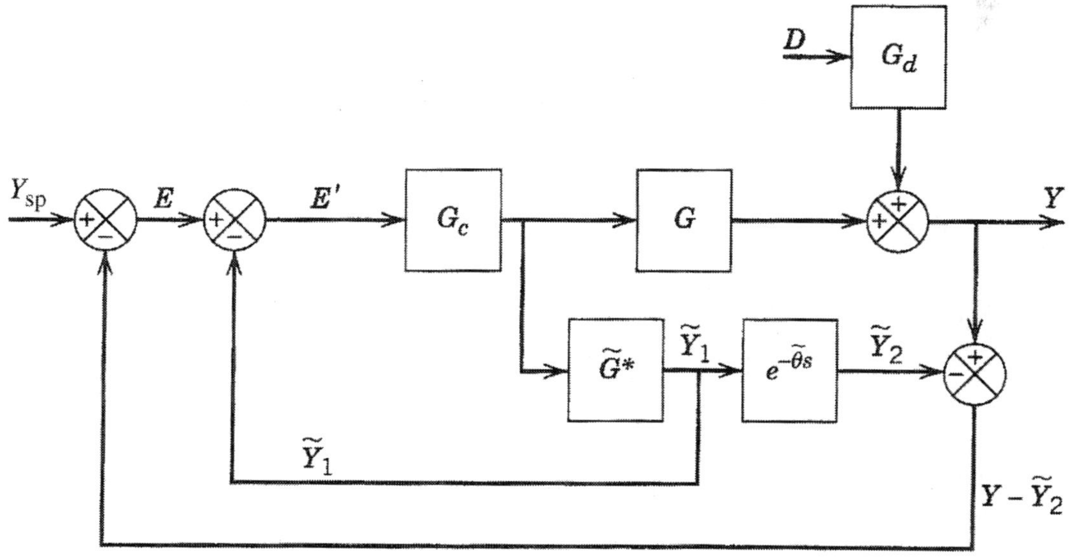

FIG. 8-38 Block diagram of the Smith predictor. (*Source: Seborg et al.*, Process Dynamics and Control, *2d ed., Wiley, New York, 2004.*)

> denotes a high selector and < a low selector. For example, a high selector can be used to determine the hot-spot temperature in a fixed-bed chemical reactor. In this case, the output from the high selector is the input to the temperature controller. In an exothermic catalytic reaction, the process may run away due to disturbances or changes in the reactor. Immediate action should be taken to prevent a dangerous rise in temperature. Because a hot spot may potentially develop at one of several possible locations in the reactor, multiple (redundant) measurement points should be employed. This approach minimizes the time required to identify when a temperature has risen too high at some point in the bed.

The use of high or low limits for process variables is another type of selective control, called an override. The feature of antireset windup in feedback controllers is a type of override. Another example is a distillation column with lower and upper limits on the heat input to the column reboiler. The minimum level ensures that liquid will remain on the trays, while the upper limit is determined by the onset of flooding. Overrides are also used in forced-draft combustion control systems to prevent an imbalance between airflow and fuel flow, which could result in unsafe operating conditions.

Other types of selective systems employ multiple final control elements or multiple controllers. In some applications, several manipulated variables are used to control a single process variable (also called split-range control). Typical examples include the adjustment of both inflow and outflow from a chemical reactor to control reactor pressure or the use of both acid and base to control pH in wastewater treatment. In this approach, the selector chooses among several controller outputs which final control element should be adjusted.

Adaptive Control Process control problems inevitably require online tuning of the controller constants to achieve a satisfactory degree of control. If the process operating conditions or the environment changes significantly, the controller may have to be retuned. If these changes occur quite frequently, then adaptive control techniques should be considered. An adaptive control system is one in which the controller parameters are adjusted automatically to compensate for changing process conditions.

The subject of adaptive control is one of current interest. New algorithms continue to be developed, but these need to be field-tested before industrial acceptance can be expected. An adaptive controller is inherently nonlinear and therefore more complicated than the conventional PID controller.

Fuzzy Logic Control The application of fuzzy logic to process control requires the concepts of fuzzy rules and fuzzy inference. A fuzzy rule, also known as a fuzzy IF-THEN statement, has the form

If x then y

if input1 = high

and input2 = low

then output = medium

Three functions are required to perform logical inferencing with fuzzy rules. The fuzzy AND is the product of a rule's input membership values, generating a weight for the rule's output. The fuzzy OR is a normalized sum of the weights assigned to each rule that contributes to a particular decision. The third function used is defuzzification, which generates a crisp final output. In one approach, the crisp output is the weighted average of the peak element values.

With a single feedback control architecture, information that is readily available to the algorithm includes the error signal, difference between the process variable and the set-point variable, change in error from previous cycles to the current cycle, changes to the set-point variable, change of the manipulated variable from cycle to cycle, and change in the process variable from past to present. In addition, multiple combinations of the system response data are available. As long as the irregularity lies in that dimension wherein fuzzy decisions are being based or associated, the result should be enhanced performance. This enhanced performance should be demonstrated in both the transient and steady-state response. If the system tends to have changing dynamic characteristics or exhibits nonlinearities, fuzzy logic control should offer a better alternative to using constant PID settings. Most fuzzy logic software begins building its information base during the autotune function. In fact, the majority of the information used in the early stages of system start-up comes from the autotune solutions.

In addition to single-loop process controllers, products that have benefited from the implementation of fuzzy logic are camcorders, elevators, antilock braking systems, and televisions with automatic color, brightness, and sound control. Sometimes fuzzy logic controllers are combined with pattern recognition software such as artificial neural networks (Passino and Yurkovich, *Fuzzy Control*, Addison-Wesley, Reading, Mass., 1998; Kosko, *Neural Networks and Fuzzy Systems*, Prentice-Hall, Englewood Cliffs, N.J., 1992).

EXPERT SYSTEMS

An expert system is a computer program that uses an expert's knowledge in a particular domain to solve a narrowly focused, complex problem. An offline system uses information entered manually and produces results in visual form to guide the user in solving the problem at hand. An online system uses information taken directly from process measurements to perform tasks automatically or instruct or alert operating personnel to the status of the plant.

Each expert system has a rule base created by the expert to respond as the expert would to sets of input information. Expert systems used for plant diagnostics and management usually have an open rule base, which can be changed and augmented as more experience accumulates and more tasks are automated. The system begins as an empty shell with an assortment of functions such as equation solving, logic, and simulation, as well as input and display tools to allow an expert to construct a proprietary rule base. The "expert" in this case would be the person or persons having the deepest knowledge about the process, its problems, its symptoms, and remedies. Converting these inputs to meaningful outputs is the principal task in constructing a rule base. First-principles models (deep knowledge) produce the most accurate results, although heuristics are always required to establish limits. Often modeling tools such as artificial neural nets are used to develop relationships among the process variables.

A number of process control vendors offer comprehensive, object-oriented software environments for building and deploying expert systems. Advantages of such software include transforming complex real-time data to useful information through knowledge-based reasoning and analysis, monitoring for potential problems before they adversely impact operations, diagnosing root causes of time-critical problems to speed up resolution, and recommending or taking corrective actions to help ensure successful recovery.

MULTIVARIABLE CONTROL

GENERAL REFERENCES: Shinskey, *Process Control Systems*, 4th ed., McGraw-Hill, New York, 1996. Seborg, Edgar, and Mellichamp, *Process Dynamics and Control*, 2d ed., Wiley, New York, 2004. McAvoy, *Interaction Analysis*, ISA, Research Triangle Park, N.C., 1983.

Process control books and journal articles tend to emphasize problems with a single controlled variable. In contrast, many processes require multivariable control with many process variables to be controlled. In fact, for virtually any important industrial process, at least two variables must be controlled: product quality and throughput. In this section, strategies for multivariable control are considered.

Three examples of simple multivariable control systems are shown in Fig. 8-39. The in-line blending system blends pure components A and B to produce a product stream with flow rate w and mass fraction of A, x. Adjusting either inlet flow rate w_A or w_B affects both of the controlled variables w and x. For the pH neutralization process in Fig. 8-39b, liquid level h and exit stream pH are to be controlled by adjusting the acid and base flow rates w_A and w_B. Each of the manipulated variables affects both of the controlled variables. Thus, both the blending system and the pH neutralization process are said to exhibit strong process interactions. In contrast, the process interactions for the gas-liquid separator in Fig. 8-39*c* are not as strong because one manipulated variable, liquid flow rate L, has only a small and indirect effect on one of the controlled variables, pressure P.

Strong process interactions can cause serious problems if a conventional multiloop feedback control scheme (e.g., either PI or PID controllers) is employed. The process interactions can produce undesirable control loop interactions where the controllers fight each other. Also, it may be difficult to determine the best pairing of controlled and manipulated variables. For example, in the in-line blending process in Fig. 8-39*a*, should w be controlled with w_A and x with w_B, or vice versa?

Control Strategies for Multivariable Control If a conventional multiloop control strategy performs poorly due to control loop interactions, a number of solutions are available:

1. Detune one or more of the control loops.
2. Choose different controlled or manipulated variables (or their pairings).
3. Use a decoupling control system.
4. Use a multivariable control scheme (e.g., model predictive control).

Detuning a controller (e.g., using a smaller controller gain or a larger reset time) tends to reduce control loop interactions by sacrificing the performance for the detuned loops. This approach may be acceptable if some of the controlled variables are faster or less important than others.

The selection of controlled and manipulated variables is of crucial importance in designing a control system. In particular, a judicious choice may significantly reduce control loop interactions. For the blending process in Fig. 8-39*a*, a straightforward control strategy would be to control x by adjusting w_A, and w by adjusting w_B. But physical intuition suggests that it would be better to control x by adjusting the ratio $w_A/(w_A + w_B)$ and to control product flow rate w by the sum $w_A + w_B$. Thus, the new manipulated variables would be $U_1 = w_A/(w_A + w_B)$ and $U_2 = w_A + w_B$. In this control scheme, U_1 affects only x, and U_2 affects only w. Thus, the control loop interactions have been eliminated. Similarly, for the pH neutralization process in Fig. 8-39*b*, the control loop interactions would be greatly reduced if pH were controlled by $U_1 = w_A\ (w_A + w_B)$ and liquid level h were controlled by $U_2 = w_A + w_B$.

Decoupling Control Systems Decoupling control systems provide an alternative approach for reducing control loop interactions. The basic idea is to use additional controllers called decouplers to compensate for undesirable process interactions.

As an illustrative example, consider the simplified block diagram for a representative decoupling control system shown in Fig. 8-40. The two controlled variables Y_1 and Y_2 and two manipulated variables U_1 and U_2 are related by the four process transfer functions G_{p11}, G_{p12}, and so on. For example, G_{p11} denotes the transfer function between U_1 and Y_1:

$$\frac{Y_1(s)}{U_1(s)} = G_{p11}(s) \tag{8-49}$$

Figure 8-40 includes two conventional feedback controllers: G_{c1} controls Y_1 by manipulating U_1, and G_{c2} controls Y_2 by manipulating U_2. The output signals from the feedback controllers serve as input signals to the two decouplers T_{12} and T_{21}. The block diagram is in a simplified form because the disturbance variables and transfer functions for the final control elements and sensors have been omitted.

The function of the decouplers is to compensate for the undesirable process interactions represented by G_{p12} and G_{p21}. Suppose that the process transfer functions are all known. Then the ideal design equations are

$$T_{12}(s) = -\frac{G_{p12}(s)}{G_{p11}(s)} \tag{8-50}$$

$$T_{21}(s) = -\frac{G_{p21}(s)}{G_{p22}(s)} \tag{8-51}$$

These decoupler design equations are very similar to the ones for feedforward control in an earlier section. In fact, decoupling can be interpreted as a type of feedforward control where the input signal is the output of a feedback controller rather than a measured disturbance variable.

In principle, ideal decoupling eliminates control loop interactions and allows the closed-loop system to behave as a set of independent control loops. But in practice, this ideal behavior is not attained for a variety of reasons, including imperfect process models and the presence of saturation constraints on controller outputs and manipulated variables. Furthermore, the ideal decoupler design equations in (8-50) and (8-51) may not be physically realizable and thus would have to be approximated.

In practice, other types of decouplers and decoupling control configurations have been employed. For example, in partial decoupling, only a single decoupler is employed (i.e., either T_{12} or T_{21} in Fig. 8-40

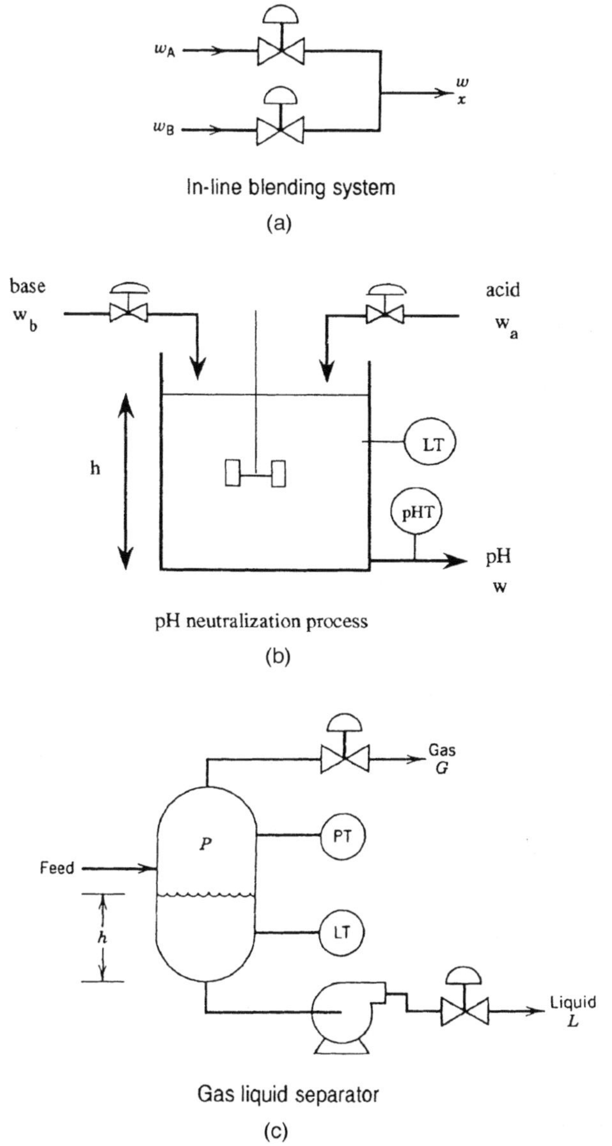

FIG. 8-39 Physical examples of multivariable control problems.

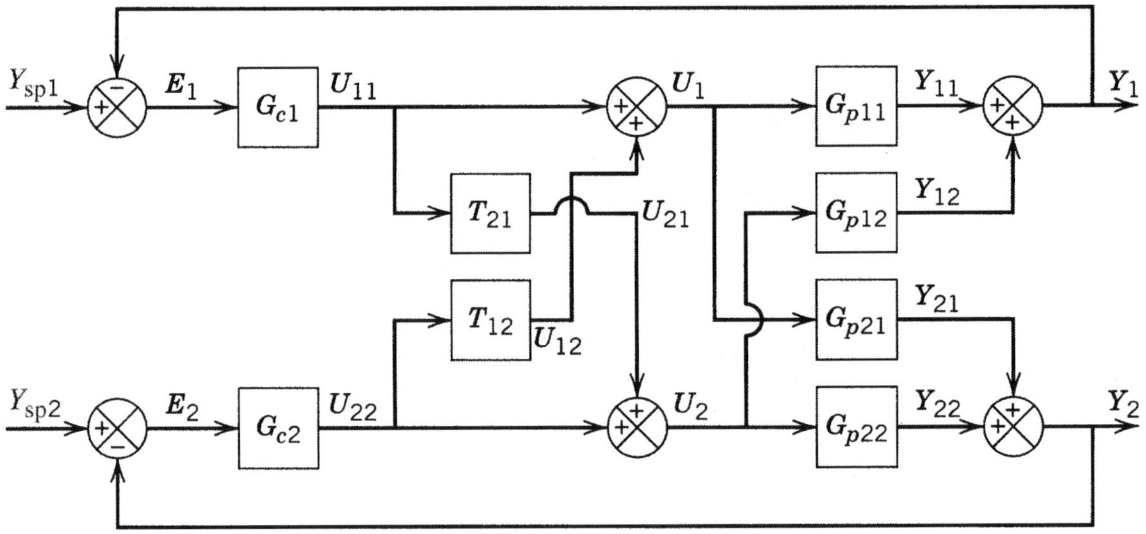

FIG. 8-40 A decoupling control system.

is set equal to zero). This approach tends to be more robust than complete decoupling and is preferred when one of the controlled variables is more important than the other. Static decouplers can be used to reduce the steady-state interactions between control loops. They can be designed by replacing the transfer functions in Eqs. (8-50) and (8-51) with the corresponding steady-state gains

$$T_{12}(s) = -\frac{K_{p12}}{K_{p11}} \tag{8-52}$$

$$T_{21}(s) = -\frac{K_{p21}}{K_{p22}} \tag{8-53}$$

The advantage of static decoupling is that less process information is required, namely, only steady-state gains. Nonlinear decouplers can be used when the process behavior is nonlinear.

Pairing of Controlled and Manipulated Variables A key decision in multiloop control system design is the pairing of manipulated and controlled variables. This is referred to as the controller pairing problem. Suppose there are N controlled variables and N manipulated variables. Then $N!$ distinct control configurations exist. For example, if $N = 5$, then there are 120 different multiloop control schemes. In practice, many would be rejected based on physical insight or previous experience. But a smaller number (say, 5 to 15) may appear to be feasible, and further analysis would be warranted. Thus, it is very useful to have a simple method for choosing the most promising control configuration.

The most popular and widely used technique for determining the best controller pairing is the relative gain array (RGA) method (Bristol, "On a New Measure of Process Interaction," *IEEE Trans. Auto. Control*, **AC-11:** 133, 1966). The RGA method provides two important items of information:

1. A measure of the degree of process interactions between the manipulated and controlled variables
2. A recommended controller pairing

An important advantage of the RGA method is that it requires minimal process information, namely, steady-state gains. Another advantage is that the results are independent of both the physical units used and the scaling of the process variables. The chief disadvantage of the RGA method is that it neglects process dynamics, which can be an important factor in the pairing decision. Thus, the RGA analysis should be supplemented with an evaluation of process dynamics. Although extensions of the RGA method that incorporate process dynamics have been reported, these extensions have not been widely applied.

RGA Method for 2 × 2 Control Problems To illustrate the use of the RGA method, consider a control problem with two inputs and two outputs. The more general case of $N \times N$ control problems is considered elsewhere (McAvoy, *Interaction Analysis*, ISA, Research Triangle Park, N.C., 1983). As a starting point, it is assumed that a linear, steady-state process model in Eqs. (8-54) and (8-55) is available, where U_1 and U_2 are steady-state values of the manipulated inputs; Y_1 and Y_2 are steady-state values of the controlled outputs; and the K values are steady-state gains. The Y and U variables are deviation variables from nominal steady-state values. This process model could be obtained in a variety of ways, such as by linearizing a theoretical model or by calculating steady-state gains from experimental data or a steady-state simulation.

$$Y_1 = K_{11}U_1 + K_{12}U_2 \tag{8-54}$$

$$Y_2 = K_{21}U_1 + K_{22}U_2 \tag{8-55}$$

By definition, the relative gain λ_{ij} between the ith manipulated variable and the jth controlled variable is defined as

$$\lambda_{ij} = \frac{\text{open-loop gain between } Y_i \text{ and } U_j}{\text{closed-loop gain between } Y_i \text{ and } U_j} \tag{8-56}$$

where the open-loop gain is simply K_{ij} from Eqs. (8-54) and (8-55). The closed-loop gain is defined to be the steady-state gain between U_j and Y_i when the other control loop is closed and no offset occurs in the other controlled variable due to the presence of integral control action. The RGA for the 2 × 2 process is denoted by

$$\Lambda = \begin{bmatrix} \lambda_{11} & \lambda_{12} \\ \lambda_{21} & \lambda_{22} \end{bmatrix} \tag{8-57}$$

The RGA has the important normalization property that the sum of the elements in each row and each column is exactly 1. Consequently, the RGA in Eq. (8-57) can be written as

$$\Lambda = \begin{bmatrix} \lambda & 1-\lambda \\ 1-\lambda & \lambda \end{bmatrix} \tag{8-58}$$

where λ can be calculated from the following formula:

$$\lambda = \frac{1}{1 - K_{12}K_{21}/(K_{11}K_{22})} \tag{8-59}$$

Ideally, the relative gains that correspond to the proposed controller pairing should have a value of 1 because Eq. (8-56) implies that the open- and closed-loop gains are then identical. If a relative gain equals 1, the steady-state operation of this loop will not be affected when the other control loop is changed from manual to automatic, or vice versa. Consequently, the recommendation for the best controller pairing is to pair the controlled and manipulated variables so that the corresponding relative gains are positive and close to 1.

RGA Example To illustrate use of the RGA method, consider the following steady-state version of a transfer function model for a pilot-scale, methanol-water distillation column (Wood and Berry, "Terminal Composition Control of a Binary Distillation Column," *Chem. Eng. Sci.*, **28:** 1707, 1973): $K_{11} = 12.8$, $K_{12} = -18.9$, $K_{21} = 6.6$, and $K_{22} = -19.4$. It follows that $\lambda = 2$ and

$$\Lambda = \begin{pmatrix} 2 & -1 \\ -1 & 2 \end{pmatrix} \tag{8-60}$$

Thus it is concluded that the column is fairly interacting and the recommended controller pairing is to pair Y_1 with U_1 and Y_2 with U_2.

MODEL PREDICTIVE CONTROL

GENERAL REFERENCES: Qin and Badgwell, *Control Eng. Practice,* **11:** 773, 2003. Rawlings, *IEEE Control Systems Magazine*, **20**(3): 38, 2000). Camacho and Bordons, *Model Predictive Control,* 2d ed., Springer-Verlag, New York, 2004. Maciejowski, *Predictive Control with Constraints*, Prentice-Hall, Upper Saddle River, N.J., 2002. Seborg, Edgar, and Mellichamp, *Process Dynamics and Control,* 2d ed., Wiley, New York, 2004, Chap. 20.

The model-based control strategy that has been most widely applied in the process industries is *model predictive control* (*MPC*). It is a general method that is especially well suited for difficult multi-input, multioutput (MIMO) control problems where there are significant interactions between the manipulated inputs and the controlled outputs. Unlike other model-based control strategies, MPC can easily accommodate inequality constraints on input and output variables such as upper and lower limits and rate-of-change limits.

A key feature of MPC is that future process behavior is predicted by using a dynamic model and available measurements. The controller outputs are calculated so as to minimize the difference between the predicted process response and the desired response. At each sampling instant, the control calculations are repeated and the predictions updated based on current measurements. In typical industrial applications, the set point and target values for the MPC calculations are updated by using online optimization based on a steady-state model of the process.

The current widespread interest in MPC techniques was initiated by pioneering research performed by two industrial groups in the 1970s. Shell Oil (Houston, Tex.) reported its Dynamic Matrix Control (DMC) approach in 1979, while a similar technique, marketed as IDCOM, was published by a small French company ADERSA in 1978. Since then, there have been thousands of applications of these and related MPC techniques in oil refineries and petrochemical plants around the world. Thus, MPC has had a substantial impact and is currently the method of choice for difficult multivariable control problems in these industries. However, relatively few applications have been reported in other process industries, even though MPC is a very general approach that is not limited to a particular industry.

Advantages and Disadvantages of MPC Model predictive control offers a number of important advantages in comparison with conventional multiloop PID control:

1. It is a general control strategy for MIMO processes with inequality constraints on input and output variables.
2. It can easily accommodate difficult or unusual dynamic behavior such as large time delays and inverse responses.
3. Because the control calculations are based on optimizing control system performance, MPC can be readily integrated with online optimization strategies to optimize plant performance.
4. The control strategy can be easily updated online to compensate for changes in process conditions, constraints, or performance criteria.

But current versions of MPC have significant disadvantages in comparison with conventional multiloop control:

1. The MPC strategy is very different from conventional multiloop control strategies and thus initially unfamiliar to plant personnel.
2. The MPC calculations can be relatively complicated [e.g., solving a linear programming (LP) or quadratic programming (QP) problem at each sampling instant] and thus require a significant amount of computer resources and effort. These optimization strategies are described in the next section.
3. The development of a dynamic model from plant data is time-consuming, typically requiring days, or even weeks, of around-the-clock plant tests.
4. Because empirical models are generally used, they are valid only over the range of conditions considered during the plant tests.
5. Theoretical studies have demonstrated that MPC can perform poorly for some types of process disturbances, especially when output constraints are employed.

Because MPC has been widely used and has had considerable impact, there is a broad consensus that its advantages far outweigh its disadvantages.

Economic Incentives for Automation Projects Industrial applications of advanced process control strategies such as MPC are motivated by the need for improvements regarding safety, product quality, environmental standards, and economic operation of the process. One view of the economics incentives for advanced automation techniques is illustrated in Fig. 8-41. Distributed control systems (DCS) are widely used for data acquisition and conventional single-loop (PID) control. The addition of advanced regulatory control systems such as selective controls, gain scheduling, and time-delay compensation can provide benefits for a modest incremental cost. But

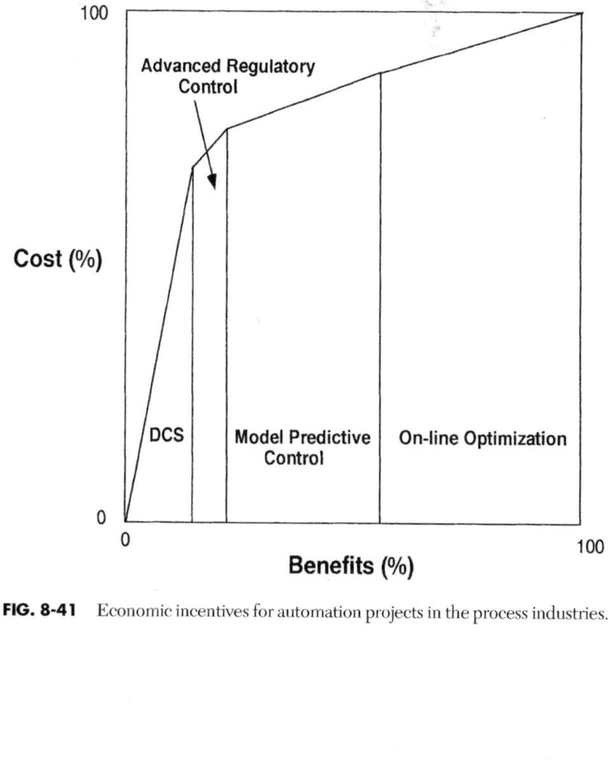

FIG. 8-41 Economic incentives for automation projects in the process industries.

experience has indicated that the major benefits can be obtained for relatively small incremental costs through a combination of MPC and online optimization. The results in Fig. 8-41 are shown qualitatively, rather than quantitatively, because the actual costs and benefits are application-dependent.

A key reason why MPC has become a major commercial and technical success is that there are numerous vendors who are licensed to market MPC products and install them on a turnkey basis. Consequently, even medium-sized companies are able to take advantage of this new technology. Payout times of 3 to 12 months have been widely reported.

Basic Features of MPC Model predictive control strategies have a number of distinguishing features:

1. A dynamic model of the process is used to predict the future outputs over a prediction *horizon* consisting of the next P sampling periods.
2. A reference trajectory is used to represent the desired output response over the prediction horizon.
3. Inequality constraints on the input and output variables can be included as an option.
4. At each sampling instant, a control policy consisting of the next M control moves is calculated. The control calculations are based on minimizing a quadratic or linear performance index over the prediction horizon while satisfying the constraints.
5. The performance index is expressed in terms of future control moves and the predicted deviations from the reference trajectory.
6. A *receding horizon approach* is employed. At each sampling instant, only the first control move (of the M moves that were calculated) is actually implemented.
7. Then the predictions and control calculations are repeated at the next sampling instant.

These distinguishing features of MPC will now be described in greater detail.

Dynamic Model A key feature of MPC is that a dynamic model of the process is used to predict future values of the controlled outputs. There is considerable flexibility concerning the choice of the dynamic model. For example, a physical model based on first principles (e.g., mass and energy balances) or an empirical model developed from data could be employed. Also, the empirical model could be a linear model (e.g., transfer function, step response model, or state space model) or a nonlinear model (e.g., neural net model). However, most industrial applications of MPC have relied on linear empirical models, which may include simple nonlinear transformations of process variables.

The original formulations of MPC (i.e., DMC and IDCOM) were based on empirical linear models expressed in either step response or impulse response form. For simplicity, we consider only a single-input, single-output (SISO) model. However, the SISO model can be easily generalized to the MIMO models that are used in industrial applications. The step response model relating a single controlled variable y and a single manipulated variable u can be expressed as

$$\hat{y}(k+1) = y_0 + \sum_{i=1}^{N-1} S_i \, \Delta u(k-i+1) + S_N u(k-N+1) \qquad (8\text{-}61)$$

where $\hat{y}(k+1)$ is the predicted value of y at the $k+1$ sampling instant, $u(k)$ is the value of the manipulated input at time k, and the model parameters S_i are referred to as the *step response coefficients*. The initial value y_0 is assumed to be known. The change in the manipulated input from one sampling instant to the next is denoted by

$$\Delta u(k) = u(k) - u(k-1) \qquad (8\text{-}62)$$

The step response model is also referred to as a discrete convolution model.

In principle, the step response coefficients can be determined from the output response to a step change in the input. A typical response to a unit step change in input u is shown in Fig. 8-42. The step response coefficients S_i are simply the values of the output variable at the sampling instants, after the initial value y_0 has been subtracted. Theoretically, they can be determined from a single step response, but, in practice, a number of "bump tests" are required to compensate for unanticipated disturbances, process nonlinearities, and noisy measurements.

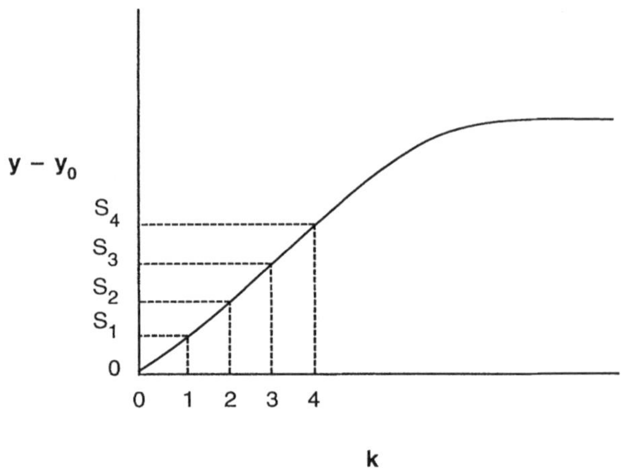

FIG. 8-42 Step response for a unit step change in the input.

Horizons The step response model in Eq. (8-61) is equivalent to the following impulse response model:

$$\hat{y}(k) = \sum_{i=1}^{N} h_i u(k-i) + y_0 \qquad (8\text{-}63)$$

where the impulse response coefficients h_i are related to the step response coefficients by $h_i = S_i - S_{i-1}$. Step and impulse response models typically contain a large number of parameters because the model horizon N is usually quite large $(30 < N < 120)$. In fact, these models are often referred to as nonparametric models.

The receding horizon feature of MPC is shown in Fig. 8-43 with the current sampling instant denoted by k. Past and present input signals $[u(i)$ for $i \le k]$ are used to predict the output at the next P sampling instants $[y(k+i)$ for $i = 1, 2, \ldots, P]$. The control calculations are performed to generate an M-step control policy $[u(k), u(k+1), \ldots, u(k+M-1)]$, which optimizes a performance index. The first control action $u(k)$ is implemented. Then at the next sampling instant $k+1$, the prediction and control calculations are repeated in order to determine $u(k+1)$. In Fig. 8-43 the reference trajectory (or target) is considered to be constant. Other possibilities include a gradual or step set-point change that can be generated by online optimization.

Performance Index The performance index for MPC applications is usually a linear or quadratic function of the predicted errors and calculated future control moves. For example, the following quadratic performance index has been widely used:

$$\min_{\Delta \mathbf{u}(k)} J = \sum_{i=1}^{P} Q_i e^2(k+i) + \sum_{i=1}^{M} R_i \, \Delta u^2(k+i-1) \qquad (8\text{-}64)$$

The value $e(k+i)$ denotes the predicted error at time $k+i$,

$$e(k+i) = y_{sp}(k+i) - \hat{y}(k+i) \qquad (8\text{-}65)$$

where $y_{sp}(k+i)$ is the set point at time $k+i$ and $\Delta\mathbf{u}(k)$ denotes the column vector of current and future control moves over the next M sampling instants:

$$\Delta\mathbf{u}(k) = [\Delta u(k), \Delta u(k+1), \ldots, \Delta u(k+M-1)]^T \qquad (8\text{-}66)$$

Equation (8-64) contains two types of design parameters that can also be used for tuning purposes. Weighting factor R_i penalizes large control moves, while weighting factor Q_i allows the predicted errors to be weighed differently at each time step, if desired.

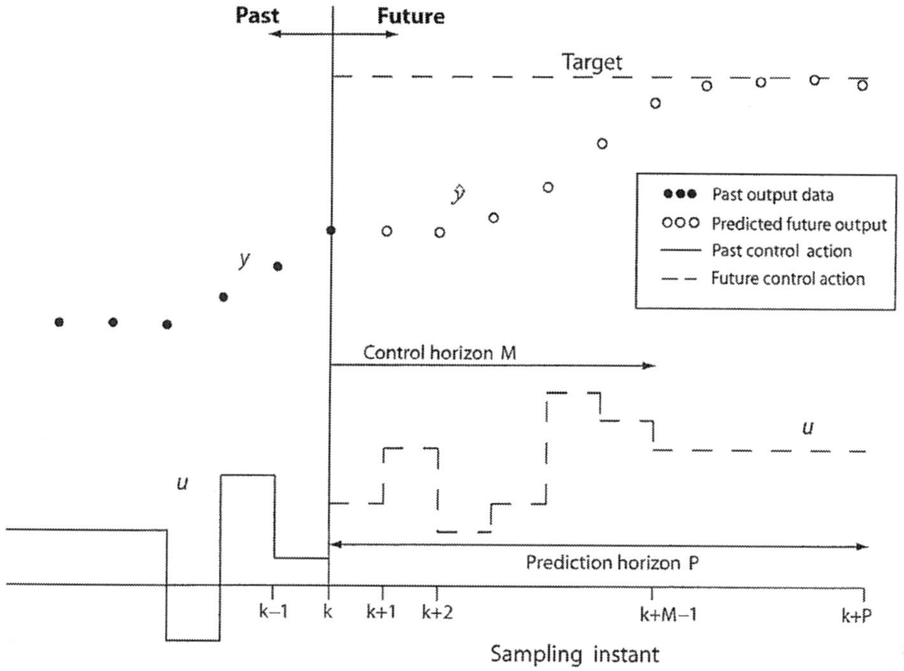

FIG. 8-43 The "moving horizon" approach of model predictive control. (*Seborg, Edgar, and Mellichamp,* Process Dynamics and Control, *2d ed., Wiley, New York, 2004.*)

Inequality Constraints Inequality constraints on the future inputs or their rates of change are widely used in the MPC calculations. For example, if both upper and lower limits are required, the constraints could be expressed as

$$u^-(k) \le u(k+j) \le u^+(k) \qquad \text{for } j = 0, 1, \ldots, M-1 \qquad (8\text{-}67)$$

$$\Delta u^-(k) \le \Delta u(k+j) \le \Delta u^+(k) \qquad \text{for } j = 0, 1, \ldots, M-1 \quad (8\text{-}68)$$

where B_i and C_i are constants. Constraints on the predicted outputs are sometimes included as well:

$$y^-(k) \le \hat{y}(k+j) \le y^+(k) \qquad \text{for } j = 0, 1, \ldots, P \qquad (8\text{-}69)$$

The minimization of the quadratic performance index in Eq. (8-64), subject to the constraints in Eqs. (8-67) to (8-69) and the step response model in Eq. (8-61), can be formulated as a standard QP (quadratic programming) problem. Consequently, efficient QP solution techniques can be employed. When the inequality constraints in Eqs. (8-67) to (8-69) are omitted, the optimization problem has an analytical solution (Camacho and Bordons, *Model Predictive Control*, 2d ed., Springer-Verlag, New York, 2004; Maciejowski, *Predictive Control with Constraints*, Prentice-Hall, Upper Saddle River, N.J., 2002). If the quadratic terms in Eq. (8-64) are replaced by linear terms, an LP (linear programming) problem results that can also be solved by using standard methods. This MPC formulation for SISO control problems can easily be extended to MIMO problems.

Implementation of MPC For a new MPC application, a cost/benefit analysis is usually performed prior to project approval. Then the steps involved in the implementation of MPC can be summarized as follows (Hokanson and Gerstle, Dynamic Matrix Control Multivariable Controllers, in *Practical Distillation Control*, Luyben (ed.), Van Nostrand Reinhold, New York, 1992, p. 248; Qin and Badgwell, *Control Eng. Practice,* **11:** 773, 2003).

Step 1: Initial Controller Design The first step in MPC design is to select the controlled, manipulated, and measured disturbance variables. These choices determine the structure of the MPC system and should be based on process knowledge and control objectives. In typical applications the number of controlled variables ranges from 5 to 40, and the number of manipulated variables is typically between 5 and 20.

Step 2: Pretest Activity During the pretest activity, the plant instrumentation is checked to ensure that it is working properly and to decide whether additional sensors should be installed. The pretest data can be used to estimate the steady-state gain and approximate settling times for each input/output pair. This information is used to plan the full plant tests of step 3.

As part of the pretest, it is desirable to benchmark the performance of the existing control system for later comparison with MPC performance (step 8).

Step 3: Plant Tests The dynamic model for the MPC calculations is developed from data collected during special plant tests. The excitation for the plant tests usually consists of changing an input variable or a disturbance variable (if possible) from one value to another, using either a series of step changes with different durations or a pseudorandom binary sequence (PRBS). To ensure that sufficient data are obtained for model identification, each input variable is typically moved to a new value, 8 to 15 times during a plant test (Qin and Badgwell, *Control Eng. Practice,* **11:** 773, 2003). Identification during closed-loop operation is becoming more common.

Step 4: Model Development The dynamic model is developed from the plant test data by selecting a model form (e.g., a step response model) and then estimating the model parameters. However, first it is important to eliminate periods of test data where plant upsets or other abnormal situations have occurred. Decisions to omit portions of the test data are based on visual inspection of the data, knowledge of the process, and experience. Parameter estimation is usually based on least squares estimation.

Step 5: Control System Design and Simulation The preliminary control system design from step 1 is critically evaluated and modified, if necessary. Then the MPC design parameters are selected including the sampling periods, weighting factors, and control and prediction horizons. Next, the closed-loop system is simulated, and the MPC design parameters are adjusted, if necessary, to obtain satisfactory control system performance and robustness over the specified range of operating conditions.

Step 6: Operator Interface Design and Operator Training Operator training is important because MPC concepts such as predictive control, multivariable interactions, and constraint handling are very different from conventional regulatory control concepts. Thus, understanding why the MPC system responds the way that it does, especially for unusual operating conditions, can be a challenge for both operators and engineers.

Step 7: Installation and Commissioning After an MPC control system is installed, it is first evaluated in a "prediction mode." Model predictions are compared with measurements, but the process continues to be controlled by the existing control system. After the output predictions are judged to be satisfactory, the calculated MPC control moves are evaluated to determine if they are reasonable. Finally, the MPC software is evaluated during closed-loop operation with the calculated control moves implemented as set points to the DCS control loops. The MPC design parameters are tuned, if necessary. The commissioning period typically requires some troubleshooting and can take as long as, or even longer than, the plant tests of step 3.

Step 8: Measuring Results and Monitoring Performance The evaluation of MPC system performance is not easy, and widely accepted metrics and monitoring strategies are not available. However, useful diagnostic information is provided by basic statistics such as the means and standard deviations for both measured variables and calculated quantities, such as control errors and model residuals. Another useful statistic is the relative amount of time that an input is saturated or a constraint is violated, expressed as a percentage of the total time the MPC system is in service.

Integration of MPC and Online Optimization As indicated in Fig. 8-41, significant potential benefits can be realized by using a combination of MPC and online optimization. At present, most commercial MPC packages integrate the two methodologies in a hierarchical configuration such as the one shown in Fig. 8-44. The MPC calculations are performed quite often (e.g., every 1 to 10 min) and implemented as set points for PID control loops at the DCS level. The targets and constraints for the MPC calculations are generated by solving a steady-state optimization problem (LP or QP) based on a linear process model. These calculations may be performed as often as the MPC calculations. As an option, the targets and constraints for the LP or QP optimization can be generated from a nonlinear process model using a nonlinear optimization technique. These calculations tend to be performed less frequently (e.g., every 1 to 24 h) due to the complexity of the calculations and the process models.

The combination of MPC and frequent online optimization has been successfully applied in oil refineries and petrochemical plants around the world.

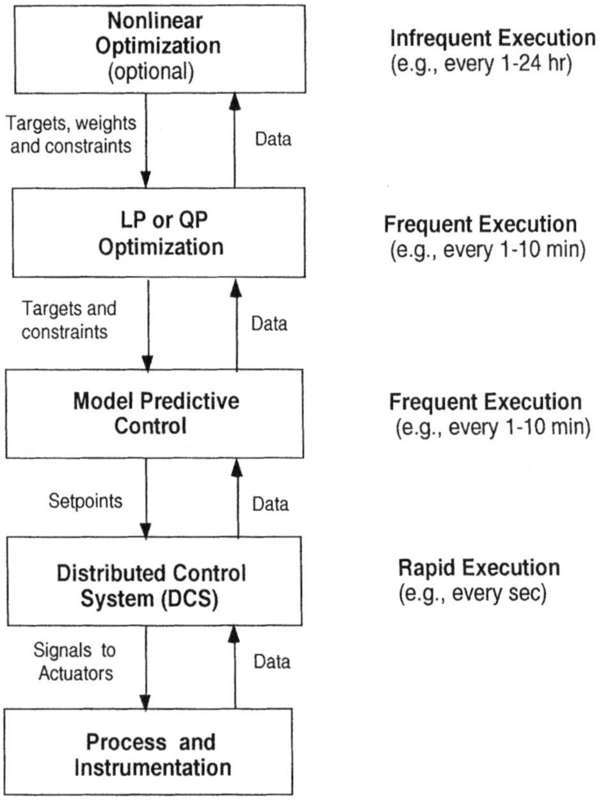

FIG. 8-44 Hierarchical control configuration for MPC and online optimization.

REAL-TIME PROCESS OPTIMIZATION

General References: Biegler, Grossmann, and Westerberg, *Systematic Methods of Chemical Process Design,* Prentice-Hall, Upper Saddle River, N.J., 1997. Darby and White, On-line Optimization of Complex Process Units, *Chem. Engr. Prog.*, **84**(10): 51, 1998. Edgar, Himmelblau, and Lasdon, *Optimization of Chemical Processes,* 2d ed., McGraw-Hill, New York, 2001. Forbes, Marlin, and MacGregor, Model Selection Criteria for Economics-Based Optimizing Control, *Comp. Chem. Engng.*, **18:** 497, 1994; Marlin and Hrymak, Real-Time Optimization of Continuous Processes, *Chem. Proc. Cont V, AIChE Symp. Ser.*, **93**(316): 156, 1997. Narashimhan and Jordache, *Data Reconciliation and Gross Error Detection*, Gulf Publishing, Houston, Tex., 2000. Nash and Sofer, *Linear and Nonlinear Programming,* McGraw-Hill, New York, 1996. Shobrys and White, Planning, Scheduling, and Control Systems: Why They Cannot Work Together, *Comp. Chem. Engng.*, **26:** 149, 2002. Timmons, Jackson, and White, Distinguishing On-line Optimization Benefits from Those of Advanced Controls, *Hydrocarb Proc.*, **79**(6): 69, 2000.

The chemical industry has undergone significant changes during the past 20 years due to the increased cost of energy and raw materials, more stringent environmental regulations, and intense worldwide competition. Modifications of both plant design procedures and plant operating conditions have been implemented to reduce costs and meet constraints. One of the most important engineering tools that can be employed in such activities is optimization. As plant computers have become more powerful, the size and complexity of problems that can be solved by optimization techniques have correspondingly expanded. A wide variety of problems in the operation and analysis of chemical plants (as well as many other industrial processes) can be solved by optimization. Real-time optimization means that the process operating conditions (set points) are evaluated on a regular basis and optimized, as shown earlier in level 4 in Fig. 8-28. Sometimes this is called steady-state optimization or supervisory control. This section examines the basic characteristics of optimization problems and their solution techniques and describes some representative benefits and applications in the chemical and petroleum industries.

Typical problems in chemical engineering process design or plant operation have many possible solutions. Optimization is concerned with selecting the best among the entire set of solutions by efficient quantitative methods. Computers and associated software make the computations involved in the selection manageable and cost-effective. Engineers work to improve the initial design of equipment and strive for enhancements in the operation of the equipment once it is installed in order to realize the greatest production, the greatest profit, the maximum cost, the least energy usage, and so on. In plant operations, benefits arise from improved plant performance, such as improved yields of valuable products (or reduced yields of contaminants), reduced energy consumption, higher processing rates, and longer times between shutdowns. Optimization can also lead to reduced maintenance costs, less equipment wear, and better staff utilization. It is helpful to systematically identify the objective, constraints, and degrees of freedom in a process or a plant if such benefits as improved quality of designs, faster and more reliable troubleshooting, and faster decision making are to be achieved.

Optimization can take place at many levels in a company, ranging from a complex combination of plants and distribution facilities down through individual plants, combinations of units, individual pieces of equipment, subsystems in a piece of equipment, or even smaller entities.

Problems that can be solved by optimization can be found at all these levels.

While process design and equipment specification are usually performed prior to the implementation of the process, optimization of operating conditions is carried out monthly, weekly, daily, hourly, or even every minute. Optimization of plant operations determines the set points for each unit at the temperatures, pressures, and flow rates that are the best in some sense. For example, the selection of the percentage of excess air in a process heater is quite critical and involves a balance on the fuel/air ratio to ensure complete combustion and at the same time make the maximum use of the heating potential of the fuel. Typical day-to-day optimization in a plant minimizes steam consumption or cooling water consumption, optimizes the reflux ratio in a distillation column, or allocates raw materials on an economic basis [Latour, *Hydro. Proc.*, **58**(6): 73, 1979, and **58**(7): 219, 1979].

A real-time optimization (RTO) system determines set-point changes and implements them via the computer control system without intervention from unit operators. The RTO system completes all data transfer, optimization calculations, and set-point implementation before unit conditions change that may invalidate the computed optimum. In addition, the RTO system should perform all tasks without upsetting plant operations. Several steps are necessary for implementation of RTO, including determination of the plant steady state, data gathering and validation, updating of model parameters (if necessary) to match current operations, calculation of the new (optimized) set points, and the implementation of these set points.

To determine if a process unit is at steady state, a program monitors key plant measurements (e.g., compositions, product rates, feed rates, and so on) and determines if the plant is close enough to steady state to start the sequence. Only when all the key measurements are within the allowable tolerances is the plant considered steady and the optimization sequence started. Tolerances for each measurement can be tuned separately. Measured data are then collected by the optimization computer. The optimization system runs a program to screen the measurements for unreasonable data (gross error detection). This validity checking automatically modifies the model updating calculation to reflect any bad data or when equipment is taken out of service. Data validation and reconciliation (online or offline) is an extremely critical part of any optimization system.

The optimization system then may run a parameter-fitting case that updates model parameters to match current plant operation. The integrated process model calculates such items as exchanger heat-transfer coefficients, reactor performance parameters, furnace efficiencies, and heat and material balances for the entire plant. Parameter fitting allows for continual updating of the model to account for plant deviations and degradation of process equipment. After completion of the parameter fitting, the information regarding the current plant constraints, control status data, and economic values for feed products, utilities, and other operating costs is collected. The economic values are updated by the planning and scheduling department on a regular basis. The optimization system then calculates the optimized set points. The steady-state condition of the plant is rechecked after the optimization case is successfully completed. If the plant is still steady, then the values of the optimization targets are transferred to the process control system for implementation. After a line-out period, the process control computer resumes the steady-state detection calculations, restarting the cycle.

Essential Features of Optimization Problems The solution of optimization problems involves the use of various tools of mathematics, which is discussed in detail in Sec. 3. The formulation of an optimization problem requires the use of mathematical expressions. From a practical viewpoint, it is important to mesh properly the problem statement with the anticipated solution technique. Every optimization problem contains three essential categories:

1. An objective function to be optimized (revenue function, cost function, etc.)
2. Equality constraints (equations)
3. Inequality constraints (inequalities)

Categories 2 and 3 comprise the model of the process or equipment; category 1 is sometimes called the economic model.

TABLE 8-6 Six Steps Used to Solve Optimization Problems

1	Analyze the process itself so that the process variables and specific characteristics of interest are defined (i.e., make a list of all the variables).
2	Determine the criterion for optimization, and specify the objective function in terms of the above variables together with coefficients. This step provides the performance model (sometimes called the economic model, when appropriate).
3	Develop via mathematical expressions a valid process or equipment model that relates the input/output variables of the process and associated coefficients. Include both equality and inequality constraints. Use well-known physical principles (mass balances, energy balances), empirical relations, implicit concepts, and external restrictions. Identify the independent and dependent variables (number of degrees of freedom).
4	If the problem formulation is too large in scope, (*a*) break it up into manageable parts and/or (*b*) simplify the objective function and model.
5	Apply a suitable optimization technique to the mathematical statement of the problem.
6	Check the answers and examine the sensitivity of the result to changes in the coefficients in the problem and the assumptions.

No single method or algorithm of optimization exists that can be applied efficiently to all problems. The method chosen for any particular case will depend primarily on (1) the character of the objective function, (2) the nature of the constraints, and (3) the number of independent and dependent variables. Table 8-6 summarizes the six general steps for the analysis and solution of optimization problems (Edgar, Himmelblau, and Lasdon, *Optimization of Chemical Processes*, 2d ed., McGraw-Hill, New York, 2001). You do not have to follow the cited order exactly, but you should cover all the steps at some level of detail. Shortcuts in the procedure are allowable, and the easy steps can be performed first. Steps 1, 2, and 3 deal with the mathematical definition of the problem: identification of variables, specification of the objective function, and statement of the constraints. If the process to be optimized is very complex, it may be necessary to reformulate the problem so that it can be solved with reasonable effort. Later in this section, we discuss the development of mathematical models for the process and the objective function (the economic model) in typical RTO applications.

Step 5 in Table 8-6 involves the computation of the optimum point. Quite a few techniques exist to obtain the optimal solution for a problem. We describe several classes of methods below. In general, the solution of most optimization problems involves the use of a digital computer to obtain numerical answers. Over the past 15 years, substantial progress has been made in developing efficient and robust computational methods for optimization. Much is known about which methods are most successful. Virtually all numerical optimization methods involve iteration, and the effectiveness of a given technique can depend on a good first guess for the values of the variables at the optimal solution. After the optimum is computed, a sensitivity analysis for the objective function value should be performed to determine the effects of errors or uncertainty in the objective function, mathematical model, or other constraints.

Development of Process (Mathematical) Models Constraints in optimization problems arise from physical bounds on the variables, empirical relations, physical laws, and so on. The mathematical relations describing the process also comprise constraints. Two general categories of models exist:

1. Those based on physical theory
2. Those based on strictly empirical descriptions

Mathematical models based on physical and chemical laws (e.g., mass and energy balances, thermodynamics, chemical reaction kinetics) are frequently employed in optimization applications. These models are conceptually attractive because a general model for any system size can be developed before the system is constructed. On the other hand, an empirical model can be devised that simply correlates input/output data without any physiochemical analysis of the process. For these models, optimization is often used to fit a model to process data, using a procedure called parameter estimation. The well-known least squares curve-fitting procedure is based on optimization theory, assuming that the model parameters are contained linearly in the model. One example is the yield matrix, where the percentage yield of each product in a unit operation is estimated for each feed component by using process data rather than employing a mechanistic set of chemical reactions.

Formulation of the Objective Function The formulation of objective functions is one of the crucial steps in the application of optimization to a practical problem. You must be able to translate the desired objective to mathematical terms. In the chemical process industries, the objective function often is expressed in units of currency per unit time (e.g., U.S. dollars per week, month, or year) because the normal industrial goal is to minimize costs or maximize profits subject to a variety of constraints.

A typical economic model involves the costs of raw materials, values of products, and costs of production as functions of operating conditions, projected sales figures, and the like. An objective function can be expressed in terms of these quantities; e.g., annual operating profit (\$/yr) might be expressed as

$$J = \sum_N F_s V_s - \sum_r F_r C_r - \text{OC} \tag{8-70}$$

where J = profit/time

$\sum_s F_s V_s$ = sum of product flow rates times respective product values (income)

$\sum_r F_r C_r$ = sum of feed flows times respective unit costs

OC = operating costs/time

Unconstrained Optimization *Unconstrained optimization* refers to the case where no inequality constraints are present and all equality constraints can be eliminated by solving for selected dependent variables followed by substitution for them in the objective function. Very few realistic problems in process optimization are unconstrained. However, the availability of efficient unconstrained optimization techniques is important because these techniques must be applied in real time, and iterative calculations may require excessive computer time. Two classes of unconstrained techniques are single-variable optimization and multivariable optimization.

Single-Variable Optimization Many real-time optimization problems can be reduced to the variation of a single-variable so as to maximize profit or some other overall process objective function. Some examples of single-variable optimization include optimizing the reflux ratio in a distillation column or the air/fuel ratio in a furnace. While most processes actually are multivariable processes with several operating degrees of freedom, often we choose to optimize only the most important variable in order to keep the strategy uncomplicated. One characteristic implicitly required in a single-variable optimization problem is that the objective function J be unimodal in the variable x.

The selection of a method for one-dimensional search is based on the tradeoff between the number of function evaluations and computer time. We can find the optimum by evaluating the objective function at many values of x, using a small grid spacing (Δx) over the allowable range of x values, but this method is generally inefficient. There are three classes of techniques that can be used efficiently for one-dimensional search: indirect, region elimination, and interpolation.

Indirect methods seek to solve the necessary condition $dJ/dx = 0$ by iteration, but these methods are not as popular as the second two classes. Region elimination methods include equal interval search, dichotomous search (or bisecting), Fibonacci search, and golden section. These methods do not use information on the shape of the function (other than its being unimodal) and thus tend to be rather conservative. The third class of techniques uses repeated polynomial fitting to predict the optimum. These interpolation methods tend to converge rapidly to the optimum without being very complicated. Two interpolation methods, quadratic and cubic interpolation, are used in many optimization packages.

Multivariable Optimization The numerical optimization of general nonlinear multivariable objective functions requires that efficient and robust techniques be employed. Efficiency is important since iteration is employed. For example, in multivariable "grid" search for a problem with four independent variables, an equally spaced grid for each variable is prescribed. For 10 values of each of the four variables, 10^4 total function evaluations would be required to find the best answer for the grid intersections, but this result may not be close enough to the true optimum and would require further search. A larger number of variables (say, 20) would require exponentially more computation, so grid search is a very inefficient method for most problems.

In multivariable optimization, the difficulty of dealing with multivariable functions is usually resolved by treating the problem as a series of one-dimensional searches. For a given starting point, a search direction s is specified, and the optimum is found by searching along that direction. The step size ε is the distance moved along s. Then a new search direction is determined, followed by another one-dimensional search. The algorithm used to specify the search direction depends on the optimization method selected.

There are two basic types of unconstrained optimization algorithms: (1) those requiring function derivatives and (2) those that do not. Here we give only an overview and refer the reader to Sec. 3 or the references for more details. The nonderivative methods are of interest in optimization applications because these methods can be readily adapted to the case in which experiments are carried out directly on the process. In such cases, an actual process measurement (such as yield) can be the objective function, and no mathematical model for the process is required. Methods that do not require derivatives are called direct methods and include sequential simplex (Nelder-Meade) and Powell's method. The sequential simplex method is quite satisfactory for optimization with two or three independent variables, is simple to understand, and is fairly easy to execute. Powell's method is more efficient than the simplex method and is based on the concept of conjugate search directions. This class of methods can be used in special cases but is not recommended for optimization involving more than 6 to 10 variables.

The second class of multivariable optimization techniques in principle requires the use of partial derivatives of the objective function, although finite difference formulas can be substituted for derivatives. Such techniques are called indirect methods and include the following classes:

1. Steepest descent (gradient) method
2. Conjugate gradient (Fletcher-Reeves) method
3. Newton's method
4. Quasi-newton methods

The steepest descent method is quite old and utilizes the intuitive concept of moving in the direction in which the objective function changes the most. However, it is clearly not as efficient as the other three. Conjugate gradient utilizes only first-derivative information, as does steepest descent, but generates improved search directions. Newton's method requires second-derivative information but is very efficient, while quasi-newton retains most of the benefits of Newton's method but utilizes only first-derivative information. All these techniques are also used with constrained optimization.

Constrained Optimization When constraints exist and cannot be eliminated in an optimization problem, more general methods must be employed than those described above, because the unconstrained optimum may correspond to unrealistic values of the operating variables. The general form of a nonlinear programming problem allows for a nonlinear objective function and nonlinear constraints, or

$$\begin{aligned} &\text{Minimize} && J(x_1, x_2, \ldots, x_n) && \\ &\text{Subject to} && h_i(x_1, x_2, \ldots, x_n) = 0 && i = 1, r_c \\ & && g_i(x_1, x_2, \ldots, x_n) > 0 && i = 1, m_c \end{aligned} \tag{8-71}$$

In this case, there are n process variables with r_c equality constraints and m_c inequality constraints. Such problems pose a serious challenge to performing optimization calculations in a reasonable amount of time. Typical constraints in chemical process optimization include operating conditions (temperatures, pressures, and flows have limits), storage capacities, and product purity specifications.

An important class of constrained optimization problems is one in which both the objective function and the constraints are linear. The solution of these problems is highly structured and can be obtained rapidly. The accepted procedure, linear programming (LP), has become quite popular in the past 20 years, solving a wide range of industrial problems. It is increasingly being used for online optimization. For processing plants, there are several different kinds of linear constraints that may arise, making the LP method of great utility.

1. Production limitation due to equipment throughput restrictions, storage limits, or market constraints
2. Raw material (feedstock) limitation
3. Safety restrictions on allowable operating temperatures and pressures
4. Physical property specifications placed on the composition of the final product. For blends of various products, we usually assume that a composite property can be calculated through the mass-averaging of pure-component physical properties
5. Material and energy balances of the steady-state model

The optimum in linear programming lies at the constraint intersections, which was generalized to any number of variables and constraints by George Dantzig. The simplex algorithm is a matrix-based numerical procedure for which many digital computer codes exist, for both mainframe and microcomputers (Edgar, Himmelblau, and Lasdon, *Optimization of Chemical Processes*, 2d ed., McGraw-Hill, New York, 2001; Nash and Sofer, *Linear and Nonlinear Programming*, McGraw-Hill, New York, 1996). The algorithm can handle virtually any number of inequality constraints and any number of variables in the objective function, and utilizes the observation that only the constraint boundaries need to be examined to find the optimum. In some instances, nonlinear optimization problems even with nonlinear constraints can be linearized so that the LP algorithm can be employed to solve them (called successive linear programming, or SLP). In the process industries, LP and SLP have been applied to a wide range of RTO problems, including refinery scheduling, olefins production, the optimal allocation of boiler fuel, and the optimization of a total plant.

Nonlinear Programming The most general case for optimization occurs when both the objective function and the constraints are nonlinear, a case referred to as nonlinear programming. While the ideas behind the search methods used for unconstrained multivariable problems are applicable, the presence of constraints complicates the solution procedure. All the methods discussed below have been utilized to solve nonlinear programming problems in the field of chemical engineering design and operations. Nonlinear programming is now used extensively in the area of real-time optimization.

One of the older and most accessible NLP algorithms uses iterative linearization and is called the *generalized reduced gradient (GRG) algorithm.* The GRG algorithm employs linear or linearized constraints and uses slack variables to convert all constraints to equality constraints. The GRG algorithm is used in the Excel Solver. CONOPT is a reduced gradient algorithm that works well for large-scale problems and nonlinear constraints. CONOPT and GRG work best for problems in which the number of degrees of freedom is small (the number of constraints is nearly equal to the number of variables).

Successive quadratic programming (*SQP*) solves a sequence of quadratic programs that approach the solution of the original NLP by linearizing the constraints and using a quadratic approximation to the objective function. Lagrange multipliers are introduced to handle constraints, and the search procedure generally employs some variation of Newton's method, a second-order method that approximates the hessian matrix using first derivatives (Biegler et al., 1997; Edgar et al., 2001). MINOS and NPSOL are software packages that are suitable for problems with large numbers of variables (more variables than equations) and constraints that are linear or nearly linear. *Successive linear programming (SLP)* is used less often for solving RTO problems. It requires linear approximations of both the objective function and the constraints but sometimes exhibits poor convergence to optima that are not located at constraint intersections.

One important class of nonlinear programming techniques is called *quadratic programming (QP),* where the objective function is quadratic and the constraints are linear. While the solution is iterative, it can be obtained quickly as in linear programming. This is the basis for the newest type of constrained multivariable control algorithms called model predictive control, which is heavily used in the refining industry. See the earlier subsection on model predictive control for more details.

Figure 8-45 gives an overview of which optimization algorithms are appropriate for certain types of RTO problems. Software libraries such as GAMS (General Algebraic Modeling System) or NAG (Numerical Algorithms Group) offer one or more NLP algorithms, but rarely are all algorithms available from a single source. Also there are quite a few good optimization software programs that are free and can be found by a web search. No single NLP algorithm is best for every problem, so several solvers should be tested on a given application. See Edgar, Himmelblau, and Lasdon, *Optimization of Chemical Processes*, McGraw-Hill, New York, 2001.

Linear and nonlinear programming solvers have been interfaced to spreadsheet software for desktop computers. The spreadsheet has become a popular user interface for entering and manipulating numeric data. Spreadsheet software increasingly incorporates analytic tools that are accessible from the spreadsheet interface and permit access to external databases. For example, Microsoft Excel incorporates an optimization-based routine called Solver that operates on the values and formulas of a spreadsheet model. Current versions (4.0 and later) include LP and NLP solvers and mixed integer programming (MIP) capability for both linear and nonlinear problems. The user specifies a set of cell addresses to be independently adjusted (the decision variables), a set of formula cells whose values are to be constrained (the constraints), and a formula cell designated as the optimization objective.

Referring to Fig. 8-30, the highest level of process control, planning and scheduling, also employs optimization extensively, often with variables that are integer. Level 5 sets production goals to meet supply and logistics constraints and addresses time-varying capacity and workforce utilization decisions. Enterprise resource planning (ERP) and supply chain management (SCM) in level 5 refer to the links in a web of relationships involving retailing (sales), distribution, transportation, and manufacturing. Planning and scheduling usually operate over relatively long time scales and tend to be decoupled from the rest of the activities in lower levels. For example, all of the refineries owned by an oil company are usually included in a comprehensive planning and scheduling model. This model can be optimized to obtain target levels and prices for interrefinery transfers, crude oil and product allocations to each refinery, production targets, inventory targets, optimal operating conditions, stream allocations, and blends for each refinery.

Some planning and scheduling problems are mixed-integer optimization problems that involve both continuous and integer problems; whether or not to operate or use a piece of equipment is a binary (on/off) decision that arises in batch processing. Solution techniques for this type of problem include branch and bound methods and global search. This latter approach handles very complex problems with multiple optima by using algorithms such as tabu search, scatter search, simulated annealing, and genetic evolutionary algorithms (see Edgar, Himmelblau, and Lasdon).

STATISTICAL PROCESS CONTROL

In industrial plants, large numbers of process variables must be maintained within specified limits in order for the plant to operate properly. Excursions of key variables beyond these limits can have significant consequences for plant safety, the environment, product quality, and plant profitability. *Statistical process control* (*SPC*), also called *statistical quality control* (*SQC*), involves the application of statistical techniques to determine whether a process is operating normally or abnormally. Thus, SPC is a process monitoring technique that relies on *quality control charts* to monitor measured variables, especially product quality.

The basic SPC concepts and control chart methodology were introduced by Walter Shewhart in the 1930s. The current widespread interest in SPC techniques began in the 1950s when they were successfully applied first in Japan and then elsewhere. Control chart methodologies are now widely used to monitor product quality and other variables that are measured infrequently or irregularly. The basic SPC methodology is described in introductory statistics textbooks (e.g., Montgomery and Runger, *Applied Statistics and Probability for Engineers,* 3d ed., Wiley, New York, 2002) and some process control textbooks (e.g., Seborg, Edgar, and Mellichamp, *Process Dynamics and Control,* 2d ed., Wiley, New York, 2004).

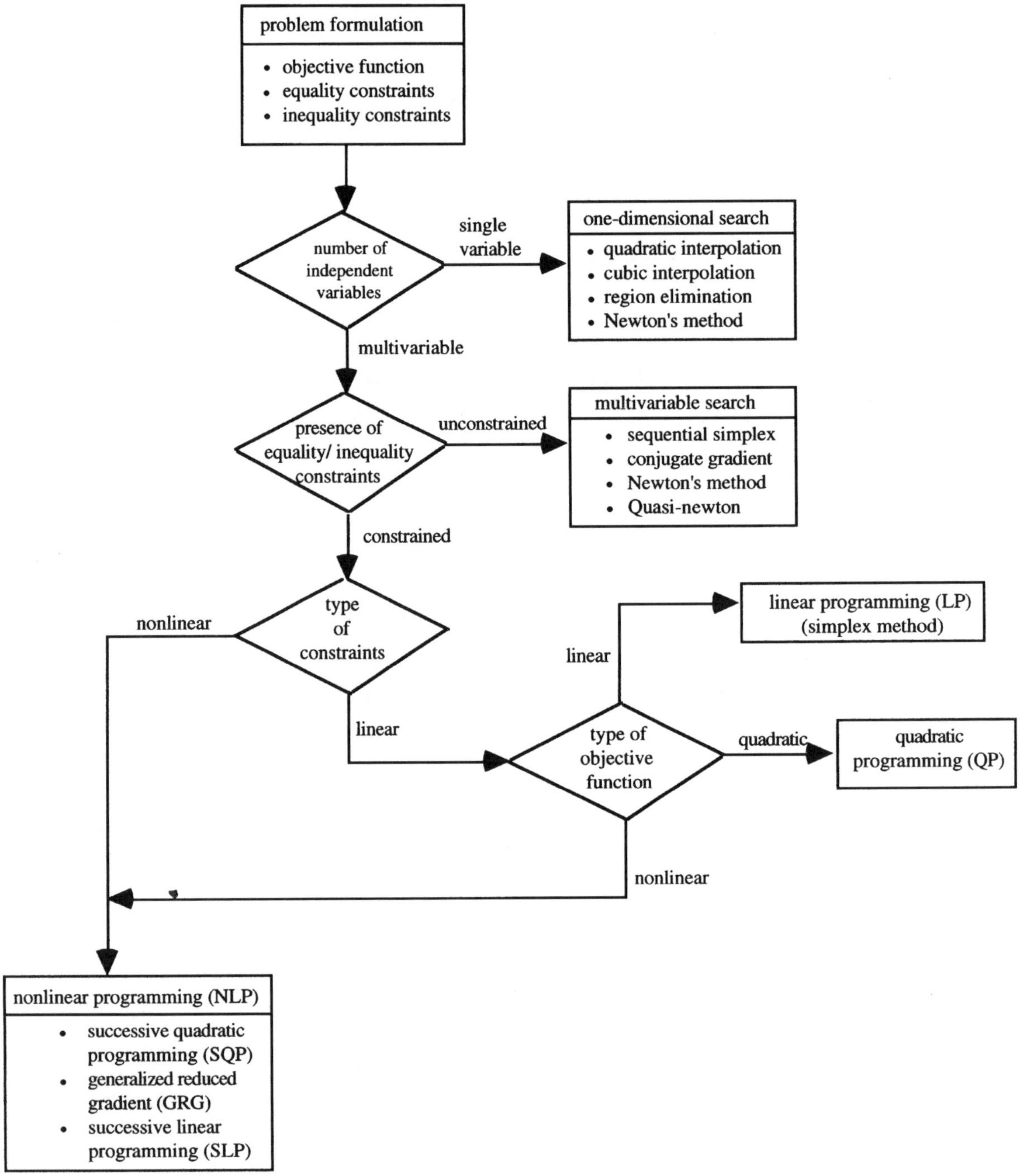

FIG. 8-45 Diagram for selection of optimization techniques with algebraic constraints and objective function.

An example of the most common control chart, the Shewhart chart, is shown in Fig. 8-46. It merely consists of measurements plotted versus sample number with control limits that indicate the range for normal process operation. The plotted data are either an individual measurement x or the sample mean $\bar{x}$ if more than one sample is measured at each sampling instant. The sample mean for k samples is calculated as,

$$\bar{x} = \frac{1}{n} \sum_{j=1}^{n} x_j \tag{8-72}$$

The Shewhart chart in Fig. 8-46 has a *target (T)*, an *upper control limit* (UCL), and a *lower control limit* (LCL). The target (or centerline) is the

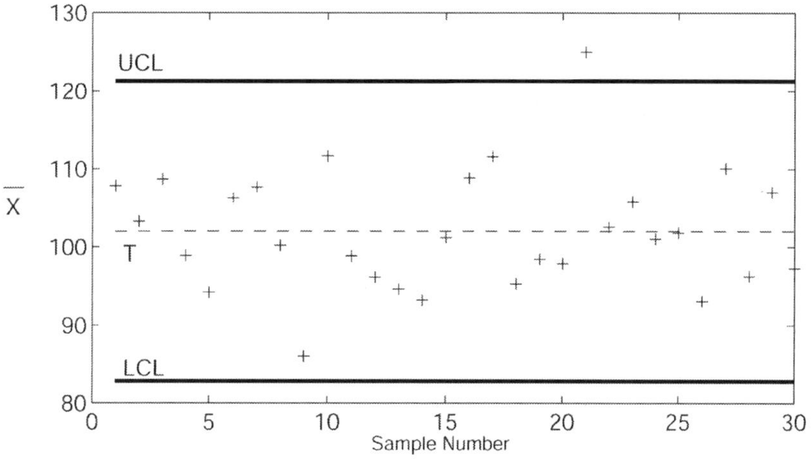

FIG. 8-46 Shewhart chart for sample mean $\bar{x}$. (*Source: Seborg et al.,* Process Dynamics and Control, *2d ed., Wiley, New York, 2004.*)

desired (or expected) value for $\bar{x}$ while the region between UCL and LCL defines the range of normal variability. If all the $\bar{x}$ data are within the control limits, the process operation is considered to be normal or "in a state of control." Data points outside the control limits are considered to be abnormal, indicating that the process operation is out of control. This situation occurs for the 21st sample. A single measurement slightly beyond a control limit is not necessarily a cause for concern. But frequent or large chart violations should be investigated to determine the root cause.

The major objective in SPC is to use process data and statistical techniques to determine whether the process operation is normal or abnormal. The SPC methodology is based on the fundamental assumption that normal process operation can be characterized by random variations around a mean value. The random variability is caused by the cumulative effects of a number of largely unavoidable phenomena such as electrical measurement noise, turbulence, and random fluctuations in feedstock or catalyst preparation. If this situation exists, the process is said to be *in a state of statistical control* (or *in control*), and the control chart measurements tend to be normally distributed about the mean value. By contrast, frequent control chart violations would indicate abnormal process behavior or an *out-of-control* situation. Then a search would be initiated to attempt to identify the *assignable cause* or the *special cause* of the abnormal behavior

The control limits in Fig. 8-46 (UCL and LCL) are based on the assumption that the measurements follow a normal distribution. Figure 8-47 shows the probability distribution for a normally distributed random variable x with mean μ and standard deviation σ. There is a very high probability (99.7 percent) that any measurement is within 3 standard deviations of the mean. Consequently, the control limits for x are typically chosen to be $T \pm 3\hat{\sigma}$, where $\hat{\sigma}$ is an estimate of σ. This estimate is usually determined from a set of representative data for a period of time when the process operation is believed to be typical. For the common situation in which the plotted variable is the sample mean, its standard deviation is estimated.

Shewhart control charts enable average process performance to be monitored, as reflected by the sample mean. It is also advantageous to monitor process variability. Process variability within a sample of k measurements can be characterized by its range, standard deviation, or sample variance. Consequently, control charts are often used for one of these three statistics.

Western Electric Rules Shewhart control charts can detect abnormal process behavior by comparing individual measurements with control chart limits. But the pattern of measurements can also provide useful information. For example, if 10 consecutive measurements are all increasing, then it is very unlikely that the process is in a

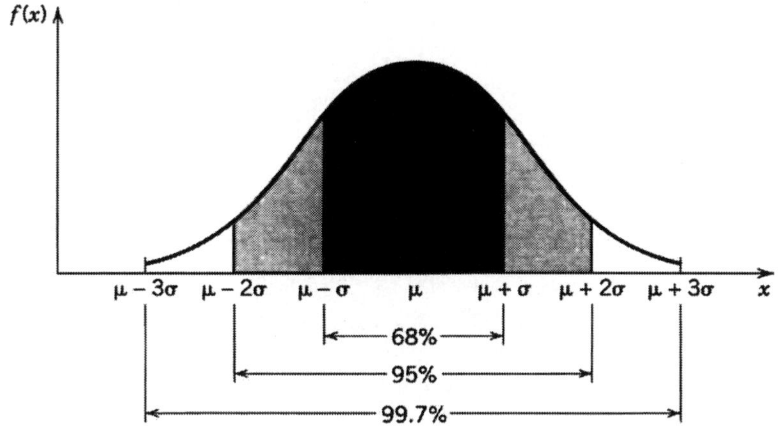

FIG. 8-47 Probabilities associated with the normal distribution. (*Source: Montgomery and Runger,* Applied Statistics and Probability for Engineers, *3d ed., Wiley, New York, 2002.*)

state of control. A wide variety of *pattern tests* (also called *zone rules*) can be developed based on the properties of the normal distribution. For example, the following excerpts from the *Western Electric Rules* (Western Electric Company, *Statistical Quality Control Handbook*, Delmar Printing Company, Charlotte, N.C., 1956; Montgomery and Runger, *Applied Statistics and Probability for Engineers,* 3d ed., Wiley, New York, 2002) indicate that the process is out of control if one or more of the following conditions occur:

1. One data point outside the 3σ control limits
2. Two out of three consecutive data points beyond a 2σ limit
3. Four out of five consecutive data points beyond a 1σ limit and on one side of the centerline
4. Eight consecutive points on one side of the centerline

Note that the first condition is the familiar Shewhart chart limits. Pattern tests can be used to augment Shewhart charts. This combination enables out-of-control behavior to be detected earlier, but the false-alarm rate is higher than that for Shewhart charts alone.

CUSUM Control Charts Although Shewhart charts with 3σ limits can quickly detect large process changes, they are ineffective for small, sustained process changes (e.g., changes smaller than 1.5σ). Alternative control charts have been developed to detect small changes such as the CUSUM control chart. They are often used in conjunction with Shewhart charts. The *cumulative sum (CUSUM)* is defined to be a running summation of the deviations of the plotted variable from its target. If the sample mean is plotted, the cumulative sum at sampling instant k, $C(k)$, is

$$C(k) = \sum_{j=1}^{k} [\bar{x}(j) - T] \tag{8-73}$$

where T is the target for $\bar{x}$. During normal process operation, $C(k)$ fluctuates about zero. But if a process change causes a small shift in $\bar{x}$, $C(k)$ will drift either upward or downward.

The CUSUM control chart was originally developed using a graphical approach based on *V masks*. However, for computer calculations, it is more convenient to use an equivalent algebraic version that consists of two recursive equations

$$C^{+}(k) = \max[0, \bar{x}(k) - (T + K) + C^{+}(k-1)] \tag{8-74}$$

$$C^{-}(k) = \max[0, (T - K) - \bar{x}(k) + C^{-}(k-1)] \tag{8-75}$$

where C^+ and C^- denote the sums for the high and low directions, and K is a constant, the *slack parameter*. The CUSUM calculations are initialized by setting $C^+(0) = C - 0 = 0$. A deviation from the target that is larger than K increases either C^+ or C^-. A control limit violation occurs when either C^+ or C^- exceeds a specified control limit (or *threshold*) H. After a limit violation occurs, that sum is reset to zero or to a specified value.

The selection of the threshold H can be based on considerations of *average run length (ARL)*, the average number of samples required to detect a disturbance of specified magnitude. For example, suppose that the objective is to be able to detect if the sample mean $\bar{x}$ has shifted from the target by a small amount δ. The slack parameter K is usually specified as $K = 0.5\delta$. For the ideal situation (e.g., normally distributed, uncorrelated disturbances), ARL values have been tabulated for different values of δ, K, and H. Table 8-7 summarizes ARL values for two values of H and different values of δ. (The values of δ are usually expressed as multiples of the standard deviation σ.) The ARL values indicate the average number of samples before a change of δ is detected. Thus the ARL values for $\delta = 0$ indicate the average time between false alarms, i.e., the average time between successive CUSUM alarms when no shift in $\bar{x}$ has occurred. Ideally, we would like the ARL value to be very large for $\delta = 0$ and small for $\delta \neq 0$. Table 8-7 shows that as the magnitude of the shift δ increases, ARL decreases and thus the CUSUM control chart detects the change faster. Increasing the value of H from 4σ to 5σ increases all the ARL values and thus provides a more conservative approach.

TABLE 8-7 Average Run Lengths for CUSUM Control Charts

Shift from target (in multiples of σ)	ARL for $H = 4\sigma$	ARL for $H = 5\sigma$
0	168.	465.
0.25	74.2	139.
0.50	26.6	38.0
0.75	13.3	17.0
1.00	8.38	10.4
2.00	3.34	4.01
3.00	2.19	2.57

Adapted from Ryan, *Statistical Methods for Quality Improvement*, 2d ed., Wiley, New York, 2000.

The relative performance of the Shewhart and CUSUM control charts is compared in Fig. 8-48 for a set of simulated data for the tensile strength of a resin. It is assumed that the tensile strength x is normally distributed with a mean of $\mu = 70$ MPa and a standard deviation of $\sigma = 3$ MPa. A single measurement is available at each sampling instant. A constant ($\sigma = 0.5\sigma = 1.5$) was added to $x(k)$ for $k \geq 10$ in order to evaluate each chart's ability to detect a small process shift. The CUSUM chart was designed using $K = 0.5\sigma$ and $H = 5\sigma$.

The Shewhart chart fails to detect the 0.5σ shift in x at $k = 10$. But the CUSUM chart quickly detects this change because a limit violation occurs at $k = 20$. The mean shift can also be detected by applying the Western Electric Rules in the previous section.

Process Capability Indices Also known as *process capability ratios*, these provide a measure of whether an in-control process is meeting its product specifications. Suppose that a quality variable x must have a volume between an *upper specification limit (USL)* and a *lower specification limit (LSL)* in order for product to satisfy customer requirements. The *capability index* C_p is defined as

$$C_p = \frac{USL - LSL}{6\sigma} \tag{8-76}$$

where σ is the standard deviation of x. Suppose that $C_p = 1$ and x is normally distributed. Based on the normal distribution, we would expect that 99.7 percent of the measurements satisfy the specification limits, or equivalently, we would expect that only 2700 out of 1 million measurements would lie outside the specification limits. If $C_p < 1$, the product specifications are satisfied; for $C_p > 1$, they are not. However, capability indices are applicable even when the data are not normally distributed.

A second capability index C_{pk} is based on average process performance $\bar{x}$ as well as process variability σ. It is defined as

$$C_{pk} = \frac{\text{Min}[\bar{x} - \text{LSL}, \text{USL} - \bar{x}]}{3\sigma} \tag{8-77}$$

Although both C_p and C_{pk} are used, we consider C_{pk} to be superior to C_p for the following reason. If $\bar{x} = T$, the process is said to be "centered" and $C_{pk} = C_p$. But for $\bar{x} \neq T$, C_p does not change, even though the process performance is worse, while C_{pk} does decrease. For this reason, C_{pk} is preferred.

If the standard deviation σ is not known, it is replaced by an estimate $\hat{\sigma}$ in Eqs. (8-76) and (8-77). For situations where there is only a single specification limit, either USL or LSL, the definitions of C_p and C_{pk} can be modified accordingly.

In practical applications, a common objective is to have a capability index of 2.0 while a value greater than 1.5 is considered to be acceptable. If the C_{pk} value is too low, it can be improved by making a change that either reduces process variability or causes $\bar{x}$ to move closer to the target. These improvements can be achieved in a number of ways that include better process control, better process maintenance, reduced variability in raw materials, improved operator training, and process changes.

Six-Sigma Approach Product quality specifications continue to become more stringent as a result of market demands and intense worldwide competition. Meeting quality requirements is especially difficult for products that consist of a very large number of components and for manufacturing processes that consist of hundreds of individual steps. For example, the production of a microelectronic device typically requires 100 to 300 batch processing steps. Suppose

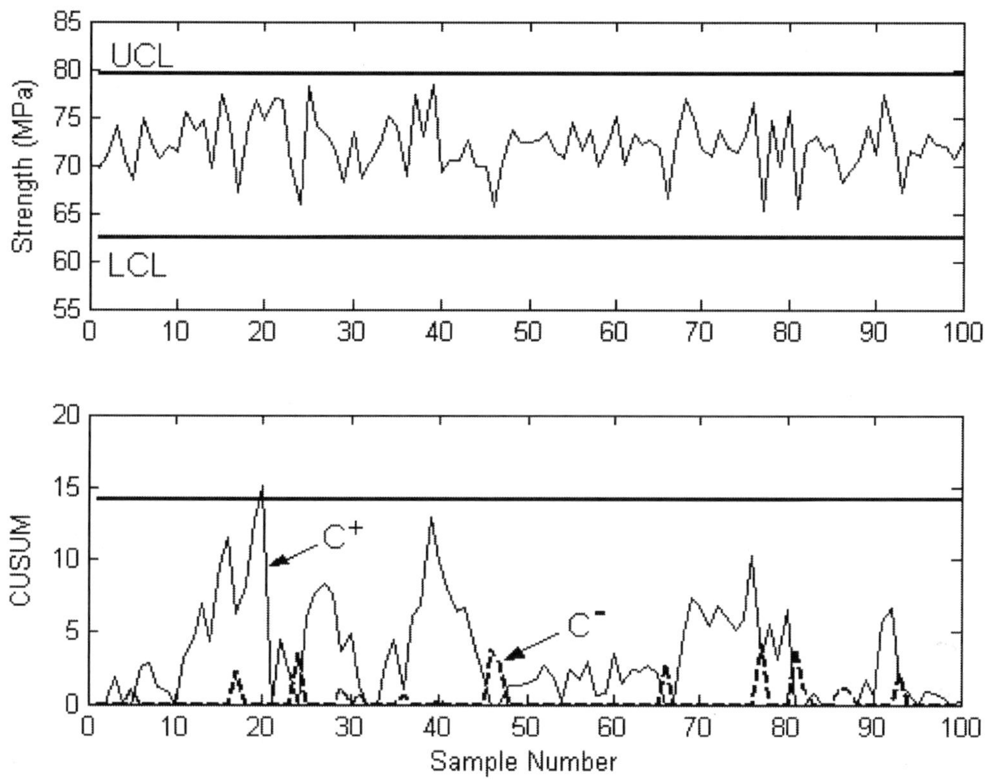

FIG. 8-48 Comparison of Shewhart and CUSUM control charts for the resin example. (*Source: Seborg et al.,* Process Dynamics and Control, *2d ed., Wiley, New York, 2004.*)

that there are 200 steps and that each one must meet a quality specification for the final product to function properly. If each step is independent of the others and has a 99 percent success rate, the overall yield of satisfactory product is $(0.99)^{200} = 0.134$, or only 13.4 percent. This low yield is clearly unsatisfactory. Similarly, even when a processing step meets 3σ specifications (99.73 percent success rate), it will still result in an average of 2700 "defects" for every 1 million produced. Furthermore, the overall yield for this 200-step process is still only 58.2 percent.

The six-sigma approach was pioneered by the Motorola Corporation in the early 1980s as a strategy for achieving both six-sigma quality and continuous improvement. Since then, other large corporations have adopted companywide programs that apply the six-sigma approach to all their business operations, both manufacturing and nonmanufacturing. Thus, although the six-sigma approach is "data-driven" and based on statistical techniques, it has evolved into a broader management philosophy that has been implemented successfully by many large corporations. The six-sigma programs have also had a significant financial impact.

Multivariate Statistical Techniques For common SPC monitoring problems, two or more quality variables are important, and they can be highly correlated. For these situations, multivariable (or *multivariate*) SPC techniques can offer significant advantages over the single-variable methods discussed earlier. Multivariate monitoring based on the classical Hotelling's T^2 statistic (Montgomery, *Introduction to Statistical Quality Control,* 4th ed., Wiley, New York, 2001) can be effective if the data are not highly correlated and the number of variables p is not large (for example, $p < 10$). Fortunately, alternative multivariate monitoring techniques such as *principal-component analysis* (*PCA*) and *partial least squares* (*PLS*) methods have been developed that are very effective for monitoring problems with large numbers of variables and highly correlated data [Piovoso and Hoo (eds.), Special Issue of *IEEE Control Systems Magazine*, **22**(5), 2002].

UNIT OPERATIONS CONTROL

PIPING AND INSTRUMENTATION DIAGRAMS

General References: Shinskey, *Process Control Systems*, 4th ed., McGraw-Hill, New York, 1996. Luyben, *Practical Distillation Control*, Van Nostrand Reinhold, New York, 1992. Luyben, Tyreus, and Luyben, *Plantwide Process Control*, McGraw-Hill, New York, 1998.

The piping and instrumentation (P&I) diagram provides a graphical representation of the control configuration of the process. P&I diagrams illustrate the measuring devices that provide inputs to the control strategy, the actuators that will implement the results of the control calculations, and the function blocks that provide the control logic. They may also include piping details such as line sizes and the location of hand valves and condensate traps.

The symbology for drawing P&I diagrams generally follows standards developed by the Instrumentation, Systems, and Automation Society (ISA). The chemicals, refining, and food industries generally follow this standard. The standards are updated from time to time, primarily because the continuing evolution in control system hardware and software provides additional capabilities for implementing

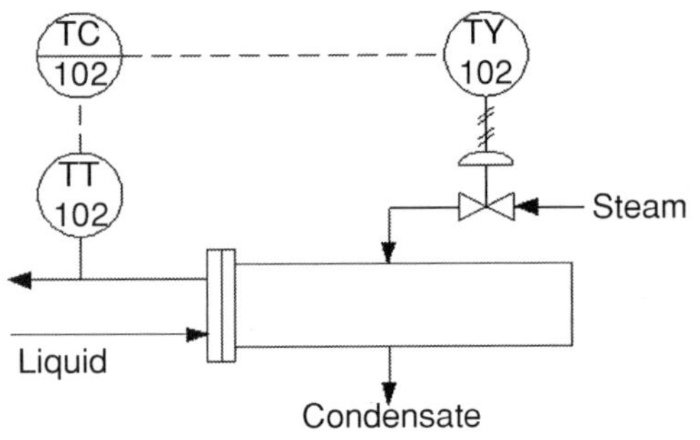

FIG. 8-49 Example of a simplified piping and instrumentation diagram.

control schemes. The ISA symbols are simple and represent a device or function as a circle containing its tag number and identifying the type of variable being controlled, e.g., pressure, and the function performed, e.g., control: PC-105. Examples of extensions of the ISA standard appear on the pages following.

Figure 8-49 presents a simplified P&I diagram for a temperature control loop that applies the ISA symbology. The measurement devices and most elements of the control logic are shown as circles:

1. TT102 is the temperature transmitter.
2. TC102 is the temperature controller.
3. TY102 is the current-to-pneumatic (I/P) transducer.

The symbol for the control valve in Fig. 8-49 is for a pneumatic modulating valve without a valve positioner.

Electronic (4- to 20-mA) signals are represented by dashed lines. In Fig. 8-49, these include the signal from the transmitter to the controller and the signal from the controller to the I/P transducer. Pneumatic signals are represented by solid lines with double crosshatching at regular intervals. The signal from the I/P transducer to the valve actuator is pneumatic.

The ISA symbology provides different symbols for different types of actuators. Furthermore, variations for the controller symbol distinguish control algorithms implemented in distributed control systems from those in panel-mounted single-loop controllers.

CONTROL OF HEAT EXCHANGERS

Steam-Heated Exchangers Steam, the most common heating medium, transfers its latent heat in condensing, causing heat flow to be proportional to steam flow. Thus a measurement of steam flow is essentially a measure of heat transfer. Consider raising a liquid from temperature T_1 to T_2 by condensing steam:

$$Q = WH = MC(T_2 - T_1) \tag{8-78}$$

where W and H are the mass flow of steam and its latent heat, M and C are the mass flow and specific heat of the liquid, and Q is the rate of heat transfer. The response of controlled temperature to steam flow is linear:

$$\frac{dT_2}{dW} = \frac{H}{MC} \tag{8-79}$$

However, the steady-state process gain described by this derivative varies inversely with liquid flow: adding a given increment of heat flow to a smaller flow of liquid produces a greater temperature rise.

Dynamically, the response of liquid temperature to a step in steam flow is that of a distributed lag, shown in Fig. 8-26 (uncontrolled). The time required to reach 63 percent complete response, $\Sigma\tau$, is essentially the residence time of the fluid in the exchanger, which is its volume divided by its flow. The residence time then varies inversely with flow. Table 8-2 gives optimum settings for PI and PID controllers for distributed lags, the proportional band varying directly with steady-state gain, and integral and derivative settings directly with $\Sigma\tau$. Since both these parameters vary inversely with liquid flow, fixed settings for the temperature controller are optimal at only one flow rate.

The variability of the process parameters with flow causes variability in load response, as shown in Fig. 8-50. The PID controller was tuned for optimum (minimum-IAE) load response at 50 percent flow. Each curve represents the response of exit temperature to a 10 percent step in liquid flow, culminating at the stated flow. The 60 percent curve is overdamped and the 40 percent curve is underdamped. The differences in gain are reflected in the amplitude of the deviation, and the differences in dynamics are reflected in the period of oscillation.

If steam flow is linear with controller output, as it is in Fig. 8-50, undamped oscillations will be produced when the flow decreases by

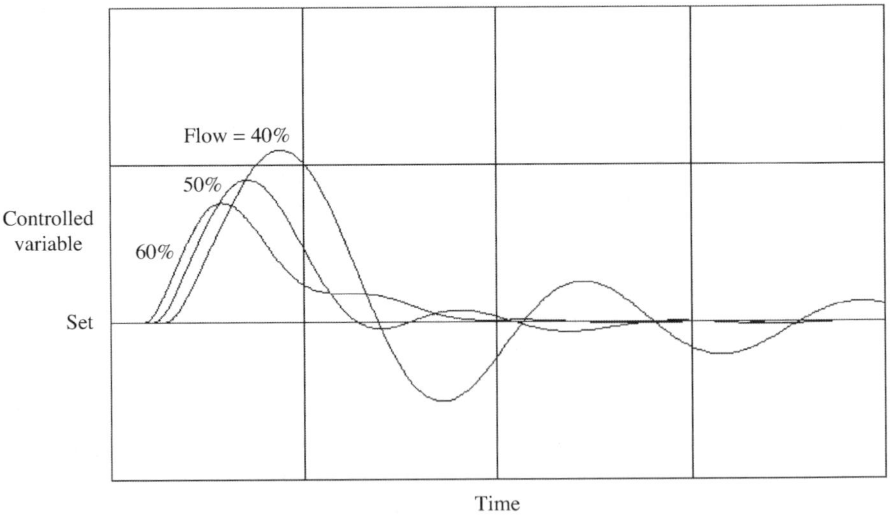

FIG. 8-50 The response of a heat exchanger varies with flow in both gain and dynamics; here the PID temperature controller was tuned for optimum response at 50 percent flow.

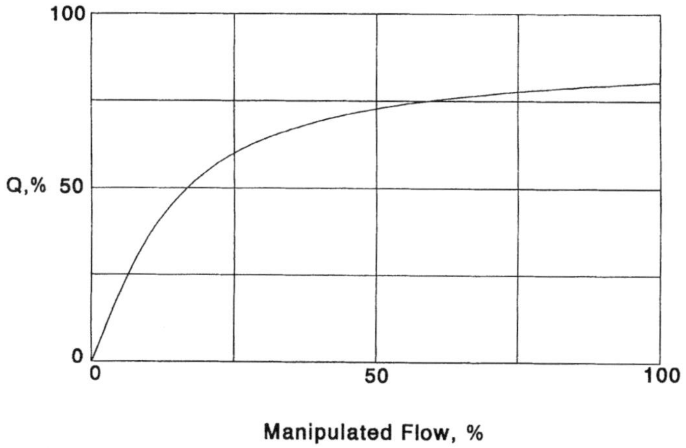

FIG. 8-51 Heat-transfer rate in sensible-heat exchange varies nonlinearly with flow of the manipulated fluid, requiring equal-percentage valve characterization.

one-third from the value at which the controller was optimally tuned—in this example at 33 percent flow. The stable operating range can be extended to one-half the original flow by using an equal-percentage (logarithmic) steam valve, whose gain varies directly with steam flow, thereby compensating for the variable process gain. Further extension requires increasing the integral setting and reducing the derivative setting from their optimum values. The best solution is to adapt all three PID settings to change inversely with measured flow, thereby keeping the controller optimally tuned for all flow rates.

Feedforward control can also be applied, as described previously under "Advanced Control Techniques." The feedforward system solves Eq. (8-78) for the manipulated set point to the steam flow controller, first by subtracting inlet temperature T_1 from the output of the outlet temperature controller (in place of T_2), and then by multiplying the result by the dynamically compensated liquid flow measurement. If the inlet temperature is not subject to rapid or wide variation, it can be left out of the calculation. Feedforward is capable of a reduction in integrated error as much as a 100-fold, but requires the use of a steam flow loop and lead-lag compensator to approach this effectiveness. Multiplication of the controller output by liquid flow varies the feedback loop gain directly proportional to flow, extending the stable operating range of the feedback loop much as the equal-percentage steam valve did without feedforward. (This system is eminently applicable to control of fired heaters in oil refineries, which commonly provide a circulating flow of hot oil to several distillation columns and are therefore subject to frequent disturbances).

Steam flow is sometimes controlled by manipulating a valve in the condensate line rather than the steam line, because it is smaller and hence less costly. Heat transfer then is changed by raising or lowering the level of condensate flooding the heat-transfer surface, an operation that is slower than manipulating a steam valve. Protection also needs to be provided against an open condensate valve blowing steam into the condensate system.

Exchange of Sensible Heat When there is no change in phase, the rate of heat transfer is no longer linear with the flow of the manipulated stream, but is a function of the mean temperature difference ΔT_m:

$$Q = UA\,\Delta T_m = M_H C_H (T_{H1} - T_{H2}) = M_C C_C (T_{C2} - T_{C1}) \quad (8\text{-}79a)$$

where U and A are the overall heat-transfer coefficient and area and subscripts H and C refer to the hot and cold fluids, respectively. An example would be a countercurrent cooler, where the hot-stream outlet temperature is controlled. Using the logarithmic mean temperature difference and solving for T_{H2} give

$$T_{H2} = T_{C1} + (T_{H1} - T_{C1}) \frac{1 - M_H C_H / M_C C_C}{\varepsilon - M_H C_H / M_C C_C} \quad (8\text{-}79b)$$

where

$$\varepsilon = \exp\left[-UA\left(\frac{1}{M_C C_C} - \frac{1}{M_H C_H}\right)\right] \quad (8\text{-}79c)$$

At a given flow of hot fluid, the heat-transfer rate is plotted as a function of coolant flow in Fig. 8-51, as a percentage of its maximum value (corresponding to $T_{C2} = T_{C1}$). The extreme nonlinearity of this relationship requires the use of an equal-percentage coolant valve for gain compensation. The variable dynamics of the distributed lag also apply, limiting the stable operating range in the same way as for the steam-heated exchanger.

Sensible-heat exchangers are also subject to variations in the temperature of the manipulated stream, an increasingly common problem where heat is being recovered at variable temperatures for reuse. Figure 8-52 shows a temperature controller (TC) setting a heat flow controller (QC) in cascade. A measurement of the manipulated flow is multiplied by its temperature difference across the heat exchanger to calculate the current heat-transfer rate, by using the right side of Eq. (8-78). Variations in supply temperature then appear as variations in calculated heat-transfer rate, which the QC can quickly correct by adjusting the manipulated flow. An equal-percentage valve is still required to linearize the secondary loop, but the primary loop of temperature-setting heat flow is linear. Feedforward can be added by multiplying the dynamically compensated flow measurement of the other fluid by the output of the temperature controller.

When a stream is manipulated whose flow is independently determined, such as the flow of a product or of a heat-transfer fluid from a fired heater, a three-way valve is used to divert the required flow to the heat exchanger. This does not alter the linearity of the process or its sensitivity to supply variations, and it even adds the possibility of independent flow variations. The three-way valve should have equal-percentage characteristics, and heat flow control may be even more beneficial.

DISTILLATION COLUMN CONTROL

Distillation columns have four or more closed loops—increasing with the number of product streams and their specifications—all of which interact with one another to some extent. Because of this interaction, there are many possible ways to pair manipulated and controlled variables through controllers and other mathematical functions, with widely differing

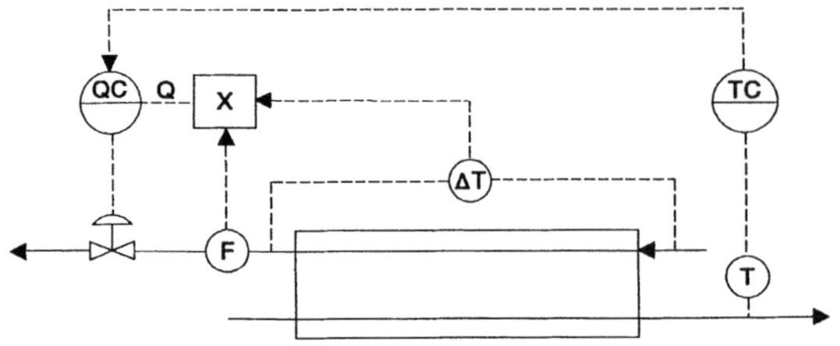

FIG. 8-52 Manipulating heat flow linearizes the loop and protects against variations in supply temperature.

degrees of effectiveness. Columns also differ from one another, so that no single rule of configuring control loops can be applied successfully to all. The following rules apply to the most common separations.

Controlling Quality of a Single Product If one of the products of a column is far more valuable than the other(s), its quality should be controlled to satisfy given specifications, and its recovery should be maximized by minimizing losses of its principal component in other streams. This is achieved by maximizing the reflux ratio consistent with flooding limits on trays, which means maximizing the flow of internal reflux or vapor, whichever is limiting. The same rule should be followed when heating and cooling have little value. A typical example is the separation of high-purity propylene from much lower-valued propane, usually achieved with the waste heat of quench water from the cracking reactors.

The most important factor affecting product quality is the material balance. In separating a feed stream F into distillate D and bottom B products, an overall mole flow balance must be maintained

$$F = D + B \tag{8-80}$$

as well as a balance on each component

$$Fz_i = Dy_i + Bx_i \tag{8-81}$$

where z, y, and x are mole fractions of component i in the respective streams. Combining these equations gives a relationship between the composition of the products and their relative portion of the feed:

$$\frac{D}{F} = 1 - \frac{B}{F} = \frac{z_i - x_i}{y_i - x_i} \tag{8-82}$$

From the above, it can be seen that control of either x_i or y_i requires both product flow rates to change with feed rate and feed composition.

Figure 8-53 shows a propylene-propane fractionator controlled at maximum boil-up by the differential pressure controller (DPC) across the trays. This loop is fast enough to reject upsets in the temperature

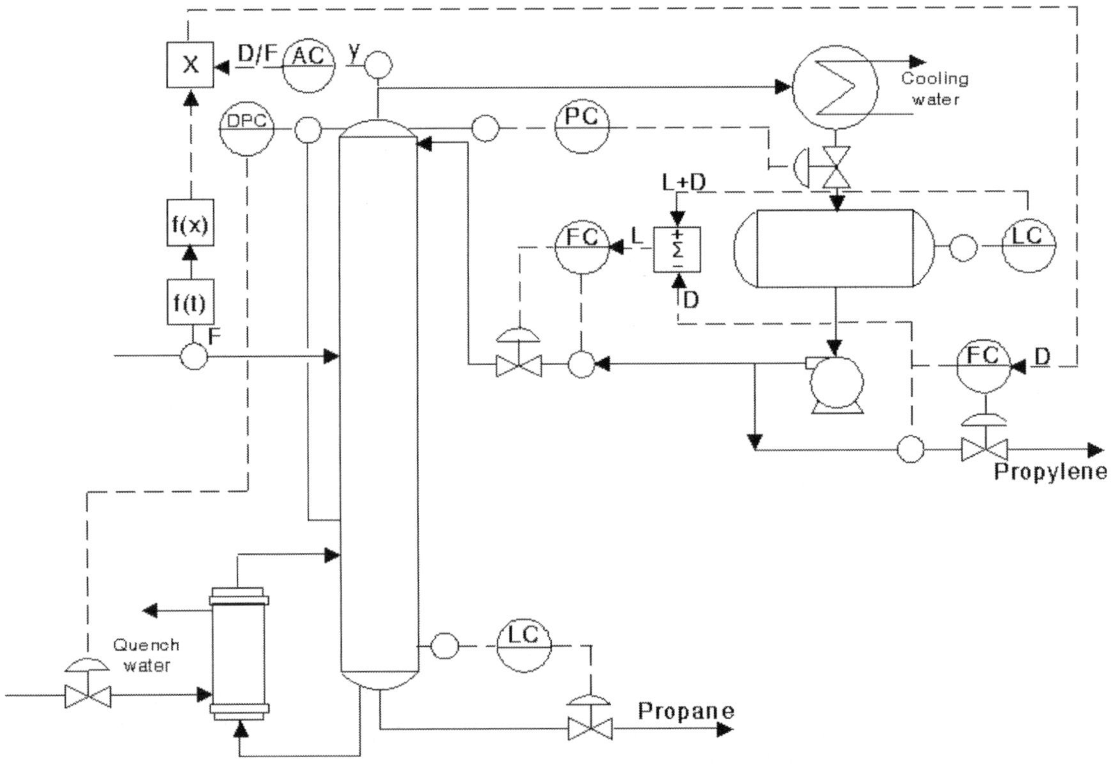

FIG. 8-53 The quality of high-purity propylene should be controlled by manipulating the material balance.

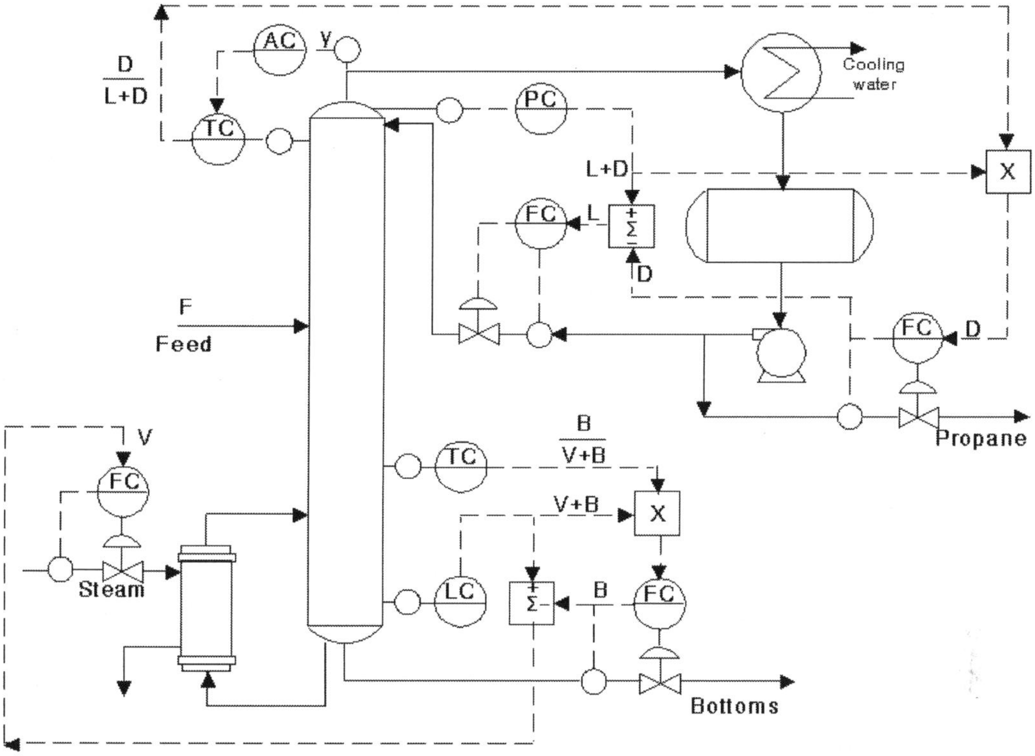

FIG. 8-54 Depropanizers require control of the quality of both products, here using reflux ratio and boil-up ratio manipulation.

of the quench water quite easily. Pressure is controlled by manipulating the heat-transfer surface in the condenser through flooding. If the condenser should become overloaded, pressure will rise above the set point, but this has no significant effect on the other control loops. Temperature measurements on this column are not helpful, as the difference between the component boiling points is too small. Propane content in the propylene distillate is measured by a chromatographic analyzer sampling the overhead vapor for fast response, and it is controlled by the analyzer controller (AC) manipulating the ratio of distillate to feed rates. The feedforward signal from feed rate is dynamically compensated by $f(t)$ and nonlinearly characterized by $f(x)$ to account for variations in propylene recovery as the feed rate changes. Distillate flow can be measured and controlled more accurately than reflux flow by a factor equal to the reflux ratio, which in this column is typically between 10 and 20. Therefore reflux flow is placed under accumulator level control (LC). Yet composition responds to the difference between internal vapor and reflux flow rates. To eliminate the lag inherent in the response of the accumulator level controller, reflux flow is driven by the subtractor in the direction opposite to distillate flow—this is essential to fast response of the composition loop. The gain of converting distillate flow changes to reflux flow changes can even be increased beyond −1, thereby changing the accumulator level loop from a lag into a dominant lead.

Controlling Quality of Two Products Where the two products have similar values, or where heating and cooling costs are comparable to product losses, the compositions of both products should be controlled. This introduces the possibility of strong interaction between the two composition loops, as they tend to have similar speeds of response. Interaction in most columns can be minimized by controlling distillate composition with reflux ratio and bottom composition with boil-up, or preferably boil-up/bottom flow ratio. These loops are insensitive to variations in feed rate, eliminating the need for feedforward control, and they also reject heat balance upsets quite effectively.

Figure 8-54 shows a depropanizer controlled by reflux and boil-up ratios. The actual mechanism through which these ratios are manipulated is as $D/(L + D)$ and $B/(V + B)$, where L is reflux flow and V is vapor boil-up, which decouples the temperature loops from the liquid-level loops. Column pressure here is controlled by flooding both the condenser and accumulator; however, there is no level controller on the accumulator, so this arrangement will not function with an overloaded condenser. Temperatures are used as indications of composition in this column because of the substantial difference in boiling points between propane and butanes. However, off-key components such as ethane do affect the accuracy of the relationship, so that an analyzer controller is used to set the top temperature controller (TC) in cascade.

If the products from a column are especially pure, even this configuration may produce excessive interaction between the composition loops. Then the composition of the less pure product should be controlled by manipulating its own flow; the composition of the remaining product should be controlled by manipulating the reflux ratio if it is the distillate, or the boil-up ratio if it is the bottom product.

Most sidestream columns have a small flow dedicated to removing an off-key impurity entering the feed, and that stream must be manipulated to control its content in the major product. For example, an ethylene fractionator separates its feed into a high-purity ethylene sidestream, an ethane-rich bottom product, and a small flow of methane overhead. This small flow must be withdrawn to control the methane content in the ethylene product. The key impurities may then be controlled in the same way as in a two-product column.

Most volatile mixtures have a relative volatility that varies inversely with column pressure. Therefore, their separation requires less energy at lower pressure, and savings in the range of 20 to 40 percent have been achieved. Column pressure can be minimized by floating on the condenser, i.e., by operating the condenser with minimal or no restrictions. In some columns, such as the propylene-propane splitter, pressure can be left uncontrolled. Where it cannot, the set point of the pressure controller can be indexed by an integral-only controller acting to slowly drive the pressure control valve toward a position just short of maximum cooling. In the case of a flooded condenser, the degree of reflux subcooling can be controlled in place of condenser valve position. Where column temperatures are used to indicate product composition, their measurements must be pressure-compensated.

CHEMICAL REACTORS

Composition Control The first requirement for successful control of a chemical reactor is to establish the proper stoichiometry, i.e., to control the flow rates of the reactants in the proportions needed to satisfy the reaction chemistry. In a continuous reactor, this begins by setting ingredient flow rates in ratio to one another. However, because of variations in the purity of the feed streams and inaccuracy in flow metering, some indication of excess reactant such as pH or a composition measurement should be used to trim the ratios. Many reactions are incomplete, leaving one or more reactants unconverted. They are separated from the products of the reaction and recycled to the reactor, usually contaminated with inert components. While reactants can be recycled to complete conversion (extinction), inerts can accumulate to the point of impeding the reaction and must be purged from the system. Inerts include noncondensible gases that must be vented and nonvolatiles from which volatile products must be stripped.

If one of the reactants differs in phase from the others and the product(s), it may be manipulated to close the material balance on that phase. For example, a gas reacting with liquids to produce a liquid product may be added as it is consumed to control reactor pressure; a gaseous purge would be necessary. Similarly, a liquid reacting with a gas to produce a gaseous product could be added as it is consumed to control the liquid level in the reactor; a liquid purge would be required. Where a large excess of one reactant *A* is used to minimize side reactions, the unreacted excess is sent to a storage tank for recycling. Its flow from the recycle storage tank is set in the desired ratio to the flow of reactant *B*, with the flow of fresh *A* manipulated to control the recycle tank level if the feed is a liquid, or tank pressure if it is a gas. Some catalysts travel with the reactants and must be recycled in the same way.

With batch reactors, it may be possible to add all reactants in their proper quantities initially, if the reaction rate can be controlled by injection of initiator or adjustment of temperature. In semibatch operation, one key ingredient is flow-controlled into the batch at a rate that sets the production. This ingredient should not be manipulated for temperature control of an exothermic reactor, as the loop includes two dominant lags—concentration of the reactant and heat capacity of the reaction mass—and can easily go unstable. It also presents the unfavorable dynamic of inverse response—increasing feed rate may lower temperature by its sensible heat before the increased reaction rate raises temperature.

Temperature Control Reactor temperature should always be controlled by heat transfer. Endothermic reactions require heat and therefore are eminently self-regulating. Exothermic reactions produce heat, which tends to raise reaction temperature, thereby increasing the reaction rate and producing more heat. This positive feedback is countered by negative feedback in the cooling system, which removes more heat as the reactor temperature rises. Most continuous reactors have enough heat-transfer surface relative to reaction mass that negative feedback dominates and they are self-regulating. But most batch reactors do not, and they are therefore steady-state unstable. Unstable reactors can be controlled if their temperature controller gain can be set high enough, and if their cooling system has enough margin to accommodate the largest expected disturbance in heat load. Stirred-tank reactors are lag-dominant, and their dynamics allow a high controller gain, but plug flow reactors are dead-time-dominant, preventing their temperature controller from providing enough gain to overcome steady-state instability. Therefore unstable plug flow reactors are also uncontrollable, their temperature tending to limit-cycle in a sawtooth wave. A stable reactor can become unstable as its heat-transfer surface fouls, or as the production rate is increased beyond a critical point (Shinskey, "Exothermic Reactors: The Stable, the Unstable, and the Uncontrollable," *Chem. Eng.*, pp. 54–59, March 2002).

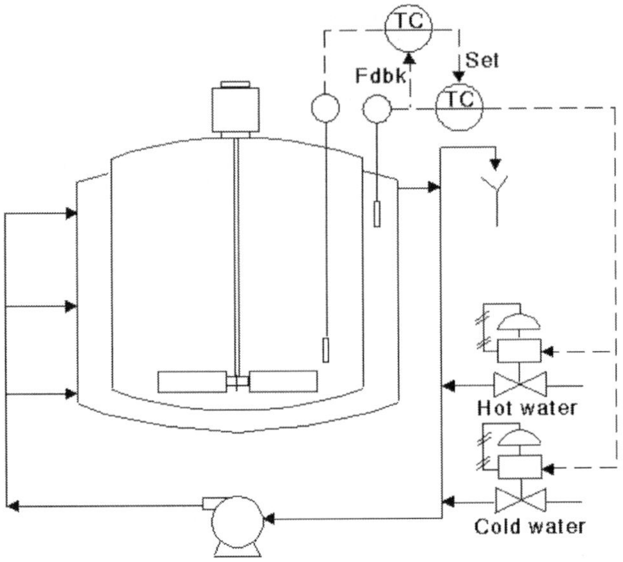

FIG. 8-55 The stirred-tank reactor temperature controller sets the coolant outlet temperature in cascade, with primary integral feedback taken from the secondary temperature measurement.

Figure 8-55 shows the recommended system for controlling the temperature of an exothermic stirred-tank reactor, either continuous or batch. The circulating pump on the coolant loop is absolutely essential to effective temperature control in keeping dead time minimum and constant—without it, dead time varies inversely with cooling load, causing limit-cycling at low loads. Heating is usually required to raise the temperature to reaction conditions, although it is often locked out of a batch reactor once the initiator is introduced. The valves are operated in split range, the heating valve opening from 50 to 100 percent of controller output and the cooling valve opening from 50 to 0 percent. The cascade system linearizes the reactor temperature loop, speeds its response, and protects it from disturbances in the cooling system. The flow of heat removed per unit of coolant flow is directly proportional to the temperature rise of the coolant, which varies with both the temperature of the reactor and the rate of heat transfer from it. Using an equal-percentage cooling valve helps compensate for this nonlinearity, but incompletely.

The flow of heat across the heat-transfer surface is linear with both temperatures, leaving the primary loop with a constant gain. Using the coolant exit temperature as the secondary controlled variable as shown in Fig. 8-55 places the jacket dynamics in the secondary loop, thereby reducing the period of the primary loop. This is dynamically advantageous for a stirred-tank reactor because of the slow response of its large heat capacity. However, a plug flow reactor cooled by an external heat exchanger lacks this heat capacity, and requires the faster response of the coolant inlet temperature loop.

Performance and robustness are both improved by using the secondary temperature measurement as the feedback signal to the integral mode of the primary controller. (This feature may be available only with controllers that integrate by positive feedback.) This places the entire secondary loop in the integral path of the primary controller, effectively pacing its integral time to the rate at which the secondary temperature is able to respond. It also permits the primary controller to be left in the automatic mode at all times without integral windup.

The primary time constant of the reactor is

$$\tau_1 = \frac{M_r C_r}{UA} \tag{8-83}$$

where M_r and C_r are the mass and heat capacity of the reactants and U and A are the overall heat-transfer coefficient and area, respectively.

The control system of Fig. 8-55 was tested on a pilot reactor where the heat-transfer area and mass could both be changed by a factor of 2, changing τ_1 by a factor of 4 as confirmed by observations of the rates of temperature rise. Yet neither controller required retuning as τ_1 varied. The primary controller should be PID and the secondary controller at least PI in this system (if the secondary controller has no integral mode, the primary will control with offset). Set-point overshoot in batch reactor control can be avoided by setting the derivative time of the primary controller higher than its integral time, but this is effective only with interacting PID controllers.

CONTROLLING EVAPORATORS

The most important consideration in controlling the quality of concentrate from an evaporator is the forcing of the vapor withdrawal rate to match the flow of excess solvent entering the feed. The mass flow rates of solid material entering and leaving are equal in the steady state

$$M_0 x_0 = M_n x_n \tag{8-84}$$

where M_0 and x_0 are the mass flow and solid fraction of the feed and M_n and x_n are their values in the product after n effects of evaporation. The total solvent evaporated from all the effects must then be

$$\sum W = M_0 - M_n = M_0\left(1 - \frac{x_0}{x_n}\right) \tag{8-85}$$

For a steam-heated evaporator, each unit of steam W_0 applied produces a known amount of evaporation, based on the number of effects and their fractional economy E:

$$\sum W = nEW_0 \tag{8-86}$$

(A comparable statement can be made with regard to the power applied to a mechanical recompression evaporator.) In summary, the steam flow required to increase the solid content of the feed from x_0 to x_n is

$$W_0 = \frac{M_0(1 - x_0/x_n)}{nE} \tag{8-87}$$

The usual measuring device for feed flow is a magnetic flowmeter, which is a volumetric device whose output F must be multiplied by density ρ to produce mass flow M_0. For most aqueous solutions fed to evaporators, the product of density and the function of solid content appearing above is linear with density:

$$F\rho\left(1 - \frac{x_0}{x_n}\right) \approx F[1 - m(\rho - 1)] \tag{8-88}$$

where slope m is determined by the desired product concentration and density is in grams per milliliter. The required steam flow in pounds per hour for feed measured in gallons per minute is then

$$W_0 = \frac{500F[1 - m(\rho - 1)]}{nE} \tag{8-89}$$

where the factor of 500 converts gallons per minute of water to pounds per hour. The factor nE is about 1.74 for a double-effect evaporator and 2.74 for a triple-effect. Using a thermocompressor (ejector) driven with 150 lb/in^2 steam on a single-effect evaporator gives an nE of 2.05; it essentially adds the equivalent of one effect to the evaporator train.

A cocurrent evaporator train with its controls is illustrated in Fig. 8-56. The control system applies equally well to countercurrent or mixed-feed evaporators, the principal difference being the tuning of the dynamic compensator $f(t)$, which must be done in the field to minimize the short-term effects of changes in feed flow on product quality. Solid concentration in the product is usually measured as density; feedback trim is applied by the analyzer controller AC adjusting the slope m of the density function, which is the only term related to x_n. This recalibrates the system whenever x_n must move to a new set point.

The accuracy of the system depends on controlling heat flow; therefore if steam pressure varies, compensation must be applied to correct

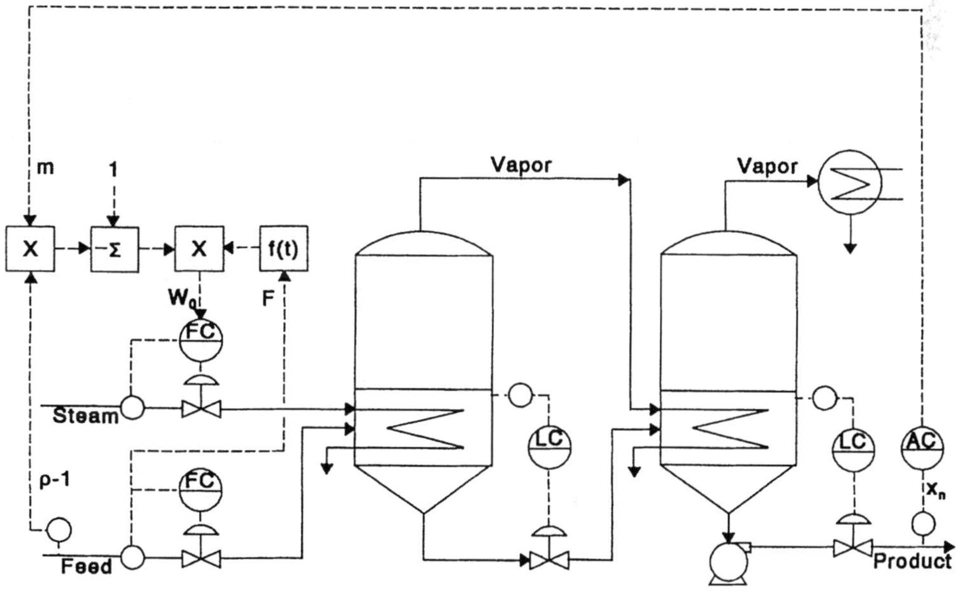

FIG. 8-56 Controlling the evaporators requires matching steam flow and evaporative load, here using feedforward control.

for both steam density and enthalpy as a function of pressure. Some evaporators must use unreliable sources of low-pressure steam. In this case, the measurement of pressure-compensated steam flow can be used to set feed flow by solving the last equation for F, using W_0 as a variable. The steam flow controller would be set for a given production rate, but the dynamically compensated steam flow measurement would be the input signal to calculate the feed flow set point. Both of these configurations are widely used in controlling corn syrup concentrators.

DRYING OPERATIONS

Controlling dryers is much different from controlling evaporators, because online measurements of feed rate and composition and product composition are rarely available. Most dryers transfer moisture from wet feed into hot dry air in a single pass. The process is generally very self-regulating, in that moisture becomes progressively harder to remove from the product as it dries: this is known as falling-rate drying. Controlling the temperature of the air leaving a cocurrent dryer tends to regulate the moisture in the product, as long as the rate and moisture content of the feed and air are reasonably constant. However, at constant outlet air temperature, product moisture tends to rise with all three of these disturbance variables.

In the absence of moisture analyzers, regulation of product quality can be improved by raising the temperature of the exhaust air in proportion to the evaporative load. The evaporative load can be estimated by the loss in temperature of the air passing through the dryer in the steady state. Changes in load are first observed in upsets in exhaust temperature at a given inlet temperature; the controller then responds by returning the exhaust air to its original temperature by changing that of the inlet air.

Figure 8-57 illustrates the simplest application of this principle as the linear relationship

$$T_0 = T_b + K\,\Delta T \tag{8-90}$$

where T_0 is the set point for exhaust temperature elevated above a base temperature T_b corresponding to zero-load operation and ΔT is the drop in air temperature from inlet to outlet. Coefficient K must be set to regulate product moisture over the expected range of evaporative load. If K is set too low, product moisture will increase with increasing load; if K is set too high, product moisture will decrease with increasing load. While K can be estimated from the model of a dryer, it does depend on the rate-of-drying curve for the product, its mean particle size, and whether the load variations are due primarily to changes in feed rate or feed moisture.

It is important to have the most accurate measurement of exhaust temperature attainable. Note that Fig. 8-57 shows the sensor inserted into the dryer upstream of the rotating seal, because air infiltration there could cause the temperature in the exhaust duct to read low—even lower than the wet-bulb temperature, an impossibility without either substantial heat loss or outside-air infiltration.

The calculation of the exhaust temperature set point forms a positive feedback loop capable of destabilizing the dryer. For example, an increase in evaporative load causes the controller to raise the inlet temperature, which will in turn raise the calculated set point, calling for a further increase in inlet temperature. The gain in the set-point loop K is typically well below the gain of the exhaust temperature measurement responding to the same change in inlet temperature. Negative feedback then dominates in the steady state, but the response of the exhaust temperature measurement is delayed by the dryer. A compensating lag $f(t)$ is shown inserted in the set-point loop to prevent positive feedback from dominating in the short term, which could cause cycling. Lag time can be safely set equal to the integral time of the outlet-air temperature controller.

If product moisture is measured offline, analytical results can be used to adjust K and T_b manually. If an online analyzer is used, the analyzer controller would be most effective in adjusting the bias T_b, as is done in the figure.

While a rotary dryer is shown, commonly used for grains and minerals, this control system has been successfully applied to fluid-bed drying of plastic pellets, air-lift drying of wood fibers, and spray drying of milk solids. The air may be steam-heated as shown or heated by direct combustion of fuel, provided that a representative measurement of inlet air temperature can be made. If it cannot, then evaporative load can be inferred from a measurement of fuel flow, which then would replace ΔT in the set-point calculation.

If the feed flows countercurrent to the air, as is the case when drying granulated sugar, the exhaust temperature does not respond to variations in product moisture. For these dryers, the moisture in the product can better be regulated by controlling its temperature at the point of discharge. Conveyor-type dryers are usually divided into a number of zones, each separately heated with recirculation of air, which raises its wet-bulb temperature. Only the last two zones may require indexing of exhaust air temperature as a function of ΔT.

Batch drying, used on small lots such as pharmaceuticals, begins operation by blowing air at constant inlet temperature through saturated product in constant-rate drying, where ΔT is constant at its maximum value ΔT_c. When product moisture reaches the point where falling-rate drying begins, the exhaust temperature begins to rise. The desired product moisture will be reached at a corresponding exhaust

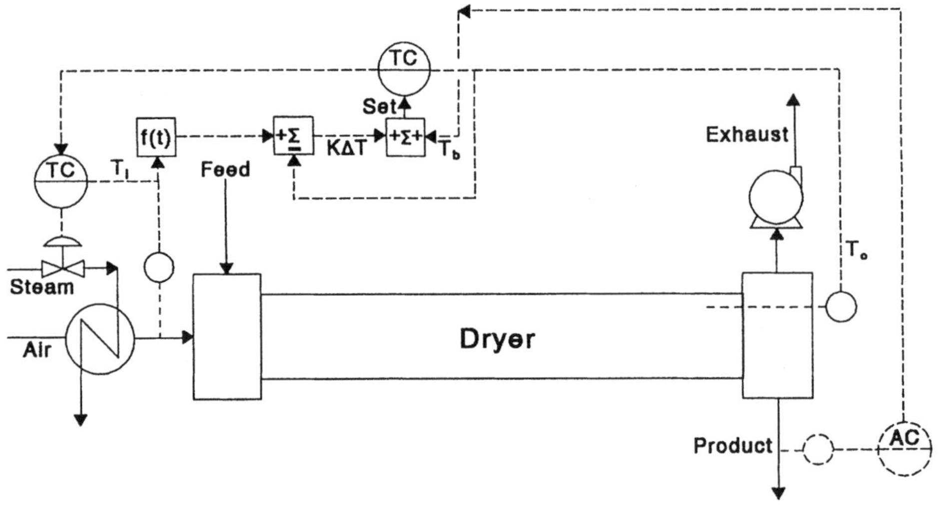

FIG. 8-57 Product moisture from a cocurrent dryer can be regulated through temperature control indexed to heat load.

temperature T_f, which is related to the temperature T_c observed during constant-rate drying, as well as to ΔT_c:

$$T_f = T_c + K \Delta T_c \tag{8-91}$$

The control system requires that the values of T_c and ΔT_c observed during the first minutes of operation be stored as the basis for the above calculation of endpoint. When the exhaust temperature then reaches the calculated value of T_f, drying is terminated. Coefficient K can be estimated from models, but requires adjustment online to reach product specifications repeatedly. Products having different moisture specifications or particle size will require different settings of K, but the system does compensate for variations in feed moisture, batch size, air moisture, and inlet temperature. Some exhaust air may be recirculated to control the dew point of the inlet air, thereby conserving energy toward the end of the batch and when the ambient air is especially dry.

BATCH PROCESS CONTROL

GENERAL REFERENCES: Fisher, *Batch Control Systems: Design, Application, and Implementation*, ISA, Research Triangle Park, N.C., 1990. Rosenof and Ghosh, *Batch Process Automation*, Van Nostrand Reinhold, New York, 1987.

BATCH VERSUS CONTINUOUS PROCESSES

When one is categorizing process plants, the following two extremes can be identified:

1. *Commodity plants*. These plants are custom-designed to produce large amounts of a single product (or a primary product plus one or more secondary products). An example is a chlorine plant, where the primary product is chlorine and the secondary products are hydrogen and sodium hydroxide. Usually the margins (product value less manufacturing costs) for the products from commodity plants are small, so the plants must be designed and operated for best possible efficiencies. Although a few are batch, most commodity plants are continuous. Factors such as energy costs are life-and-death issues for such plants.

2. *Specialty plants*. These plants are capable of producing small amounts of a variety of products. Such plants are common in fine chemicals, pharmaceuticals, foods, and so on. In specialty plants, the margins are usually high, so factors such as energy costs are important but not life-and-death issues. As the production amounts are relatively small, it is not economically feasible to dedicate processing equipment to the manufacture of only one product. Instead, batch processing is utilized so that several products (perhaps hundreds) can be manufactured with the same process equipment. The key issue in such plants is to manufacture consistently each product in accordance with its specifications.

The above two categories represent the extremes in process configurations. The term *semibatch* designates plants in which some processing is continuous but other processing is batch. Even processes that are considered to be continuous can have a modest amount of batch processing. For example, the reformer unit within a refinery is thought of as a continuous process, but the catalyst regeneration is normally a batch process.

In a continuous process, the conditions within the process are largely the same from one day to the next. Variations in feed composition, plant utilities (e.g., cooling water temperature), catalyst activities, and other variables occur, but normally these changes either are about an average (e.g., feed compositions) or exhibit a gradual change over an extended period (e.g., catalyst activities). Summary data such as hourly averages, daily averages, and the like are meaningful in a continuous process.

In a batch process, the conditions within the process are continually changing. The technology for making a given product is contained in the product recipe that is specific to that product. Such recipes normally state the following:

1. *Raw material amounts*. This is the stuff needed to make the product.

2. *Processing instructions*. This is what must be done with the stuff to make the desired product.

This concept of a recipe is quite consistent with the recipes found in cookbooks. Sometimes the term *recipe* is used to designate only the raw material amounts and other parameters to be used in manufacturing a batch. Although appropriate for some batch processes, this concept is far too restrictive for others. For some products, the differences from one product to the next are largely physical as opposed to chemical. For such products, the processing instructions are especially important. The term *formula* is more appropriate for the raw material amounts and other parameters, with *recipe* designating the formula and the processing instructions. The above concept of a recipe permits the following three different categories of batch processes to be identified:

1. *Cyclical batch*. Both the formula and the processing instructions are the same from batch to batch. Batch operations within processes that are primarily continuous often fall into this category. The catalyst regenerator within a reformer unit is a cyclical batch process.

2. *Multigrade*. The processing instructions are the same from batch to batch, but the formula can be changed to produce modest variations in the product. In a batch PVC plant, the different grades of PVC are manufactured by changing the formula. In a batch pulp digester, the processing of each batch or cook is the same, but at the start of each cook, the process operator is permitted to change the formula values for chemical-to-wood ratios, cook time, cook temperature, and so on.

3. *Flexible batch*. Both the formula and the processing instructions can change from batch to batch. Emulsion polymerization reactors are a good example of a flexible batch facility. The recipe for each product must detail both the raw materials required and how conditions within the reactor must be sequenced to make the desired product.

Of these, the flexible batch is by far the most difficult to automate and requires a far more sophisticated control system than either the cyclical batch or the multigrade batch facility.

Batches and Recipes Each batch of product is manufactured in accordance with a product recipe, which contains all information (formula and processing instructions) required to make a batch of the product (see Fig. 8-58). For each batch of product, there will be one and only one product recipe. However, a given product recipe is normally used to make several batches of product. To uniquely identify a batch of product, each batch is assigned a unique identifier called the batch ID. Most companies adopt a convention for generating the batch ID, but this convention varies from one company to the next. In most batch facilities, more than one batch of product will be in some stage of production at any given time. The batches in progress may or may not be using the same recipe. The maximum number of batches that can be in progress at any given time is a function of the equipment configuration for the plant.

The existence of multiple batches in progress at a given time presents numerous opportunities for the process operator to make errors, such as charging a material to the wrong batch. Charging a material to the wrong batch is almost always detrimental to the batch to which the material is incorrectly charged. Unless this error is recognized quickly so that the proper charge can be made, the error is also detrimental to the batch to which the charge was supposed to have been made. Such errors usually lead to an off-specification batch, but the consequences could be more serious and could result in a hazardous condition.

Recipe management refers to the assumption of such duties by the control system. Each batch of product is tracked throughout its production, which may involve multiple processing operations on various pieces of processing equipment. Recipe management ensures that all actions specified in the product recipe are performed on each batch of product made in accordance with that recipe. As the batch proceeds from one piece of processing equipment to the next, recipe management

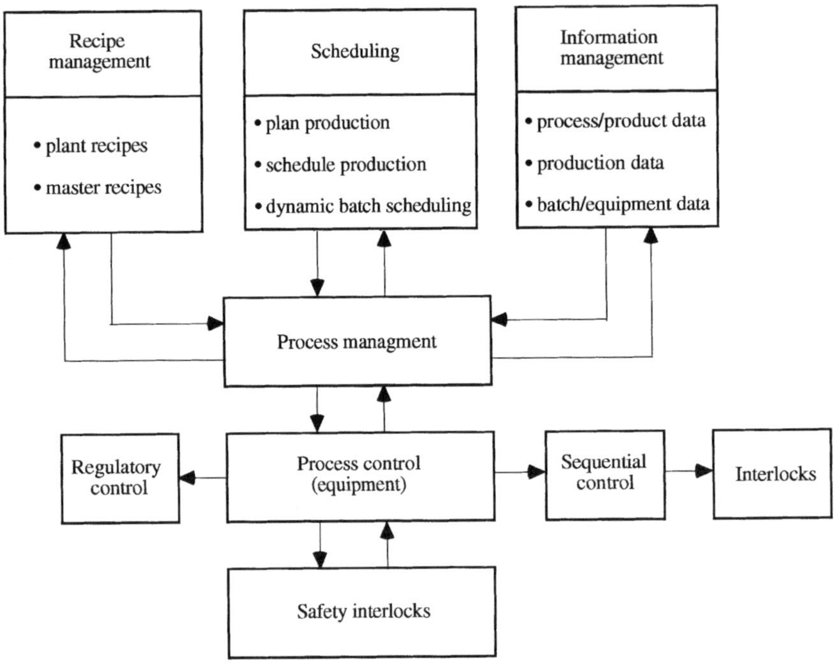

FIG. 8-58 Batch control overview.

is also responsible for ensuring that the proper type of process equipment is used and that this processing equipment is not currently in use by another batch.

By assuming such responsibilities, the control system greatly reduces the incidences where operator error results in off-specification batches. Such a reduction in error is essential to implement just-in-time production practices, where each batch of product is manufactured at the last possible moment. When a batch (or batches) is made today for shipment by overnight truck, there is insufficient time for producing another batch to make up for an off-specification batch.

Routing and Production Monitoring In some facilities, batches are individually scheduled. However, in most facilities, production is scheduled by product runs (also called process orders), where a run is the production of a stated quantity of a given product. From the stated quantity and the standard yield of each batch, the number of batches can be determined. As this is normally more than one batch of product, a production run is normally a sequence of some number of batches of the same product.

In executing a production run, the following issues must be addressed (see Fig. 8-58):

1. *Processing equipment must be dedicated to making the run.* More than one run is normally in progress at a given time. The maximum number of runs simultaneously in progress depends on the equipment configuration of the plant. Routing involves determining which processing equipment will be used for each production run.

2. *Raw material must be utilized.* When a production run is scheduled, the necessary raw materials must be allocated to the production run. As the individual batches proceed, the consumption of raw materials must be monitored for consistency with the allocation of raw materials to the production run.

3. *The production quantity for the run must be achieved by executing the appropriate number of batches.* The number of batches is determined from a standard yield for each batch. However, some batches may achieve yields higher than the standard yield, but other batches may achieve yields lower than the standard yield. The actual yields from each batch must be monitored, and significant deviations from the expected yields must be communicated to those responsible for scheduling production.

The last two activities are key components of production monitoring, although production monitoring may also involve other activities such as tracking equipment utilization.

Production Scheduling In this regard, it is important to distinguish between scheduling runs (sometimes called long-term scheduling) and assigning equipment to runs (sometimes called routing or short-term scheduling). As used here production scheduling refers to scheduling runs and is usually a corporate-level as opposed to a plant-level function. Short-term scheduling or routing was previously discussed and is implemented at the plant level. The long-term scheduling is basically a material resources planning (MRP) activity involving the following:

1. *Forecasting.* Orders for long-delivery raw materials are issued at the corporate level based on the forecast for the demand for products. The current inventory of such raw materials is also maintained at the corporate level. This constitutes the resources from which products can be manufactured. Functions of this type are now incorporated into supply chain management.

2. *Orders for products.* Orders are normally received at the corporate level and then assigned to individual plants for production and shipment. Although the scheduling of some products is based on required product inventory levels, scheduling based on orders and shipping directly to the customer (usually referred to as just-in-time) avoids the costs associated with maintaining product inventories.

3. *Plant locations and capacities.* While producing a product at the nearest plant usually lowers transportation costs, plant capacity limitations sometimes dictate otherwise. Any company competing in the world economy needs the flexibility to accept orders on a worldwide basis and then assign them to individual plants to be filled. Such a function is logically implemented within the corporate-level information technology framework.

BATCH AUTOMATION FUNCTIONS

Automating a batch facility requires a spectrum of functions.

Interlocks Some of these are provided for safety and are properly called safety interlocks. However, others are provided to avoid mistakes in processing the batch. When safety is not involved, terms such as *permissives* and *process actions* are sometimes used in lieu of interlocks. Some understand the term *interlock* to have a connection to safety (interlock will be subsequently defined as a protective response initiated on the detection of a process hazard).

Discrete Device States Discrete devices such as two-position valves can be driven to either of two possible states. Such devices can be optionally outfitted with limit switches that indicate the state of the device. For two-position valves, the following combinations are possible:

1. No limit switches
2. One limit switch on the closed position
3. One limit switch on the open position
4. Two limit switches

In process control terminology, the discrete device driver is the software routine that generates the output to a discrete device such as a valve and also monitors the state feedback information to ascertain that the discrete device actually attains the desired state. Given the variety of discrete devices used in batch facilities, this logic must include a variety of capabilities. For example, valves do not instantly change states; instead each valve exhibits a travel time for the change from one state to another. To accommodate this characteristic of the field device, the processing logic within the discrete device driver must provide for a user-specified transition time for each field device. When equipped with limit switches, the potential states for a valve are as follows:

1. *Open.* The valve has been commanded to open, and the limit switch inputs are consistent with the open state.
2. *Closed.* The valve has been commanded to close, and the limit switch inputs are consistent with the closed state.
3. *Transition.* This is a temporary state that is only possible after the valve has been commanded to change state. The limit switch inputs are not consistent with the commanded state, but the transition time has not expired.
4. *Invalid.* The transition time has expired, and the limit switch inputs are not consistent with the commanded state for the valve.

The invalid state is an abnormal condition that is generally handled in a manner similar to process alarms. The transition state is not considered to be an abnormal state but may be implemented in either of the following ways:

1. *Drive and wait.* Further actions are delayed until the device attains its commanded state.
2. *Drive and proceed.* Further actions are initiated while the device is in the transition state.

The latter is generally necessary for devices with long travel times, such as flush-fitting reactor discharge valves that are motor-driven. Closing of such valves is normally done via drive and wait; however, drive and proceed is usually appropriate when opening the valve. Although two-state devices are most common, the need occasionally arises for devices with three or more states. For example, an agitator may be on high speed, on slow speed, or off.

Process States Batch processing usually involves imposing the proper sequence of states on the process. For example, a simple blending sequence might be as follows:

1. Transfer specified amount of material from tank A to tank R. The process state is "transfer from A."
2. Transfer specified amount of material from tank B to tank R. The process state is "transfer from B."
3. Agitate for specified time. The process state is "agitate without cooling."
4. Cool (with agitation) to specified target temperature. The process state is "agitate with cooling."

For each process state, the various discrete devices are expected to be in a specified device state. For process state "transfer from A," the device states might be as follows:

1. Tank A discharge valve: open
2. Tank R inlet valve: open
3. Tank A transfer pump: running
4. Tank R agitator: off
5. Tank R cooling valve: closed

For many batch processes, process state representations are a very convenient mechanism for representing the batch logic. A grid or table can be constructed, with the process states as rows and the discrete device states as columns (or vice versa). For each process state, the state of every discrete device is specified to be one of the following:

1. Device state 0, which may be valve closed, agitator off, and so on
2. Device state 1, which may be valve open, agitator on, and so on
3. No change or don't care

This representation is easily understandable by those knowledgeable about the process technology and is a convenient mechanism for conveying the process requirements to the control engineers responsible for implementing the batch logic.

Many batch software packages also recognize process states. A configuration tool is provided to define a process state. With such a mechanism, the batch logic does not need to drive individual devices but can simply command that the desired process state be achieved. The system software then drives the discrete devices to the device states required for the target process state. This normally includes the following:

1. Generating the necessary commands to drive each device to its proper state.
2. Monitoring the transition status of each device to determine when all devices have attained their proper states.
3. Continuing to monitor the state of each device to ensure that the devices remain in their proper states. Should any discrete device not remain in its target state, failure logic must be initiated.

Regulatory Control For most batch processes, the discrete logic requirements overshadow the continuous control requirements. For many batch processes, the continuous control can be provided by simple loops for flow, pressure, level, and temperature. However, very sophisticated advanced control techniques are occasionally applied. As temperature control is especially critical in reactors, the simple feedback approach is replaced by model-based strategies that rival, if not exceed, the sophistication of advanced control loops in continuous plants.

In some installations, alternative approaches for regulatory control may be required. Where a variety of products are manufactured, the reactor may be equipped with alternative heat removal capabilities, including the following:

1. Jacket filled with cooling water. Most such jackets are once-through, but some are recirculating.
2. Heat exchanger in a pump-around loop.
3. Reflux condenser.

The heat removal capability to be used usually depends on the product being manufactured. Therefore, regulatory loops must be configured for each possible option, and sometimes for certain combinations of the possible options. These loops are enabled and disabled depending on the product being manufactured.

The interface between continuous controls and sequence logic (discussed shortly) is also important. For example, a feed might be metered into a reactor at a variable rate, depending on another feed or possibly on reactor temperature. However, the product recipe calls for a specified quantity of this feed. The flow must be totalized (i.e., integrated), and when the flow total attains a specified value, the feed must be terminated. The sequence logic must have access to operational parameters such as controller modes. That is, the sequence logic must be able to switch a controller to manual, automatic, or cascade. Furthermore, the sequence logic must be able to force the controller output to a specified value.

Sequence Logic Sequence logic must not be confused with discrete logic. Discrete logic is especially suitable for interlocks or permissives; e.g., the reactor discharge valve must be closed for the feed valve to be opened. Sequence logic is used to force the process to attain the proper sequence of states. For example, a feed preparation might be to first charge A, then charge B, next mix, and finally cool. Although discrete logic can be used to implement sequence logic, other alternatives are often more attractive.

Sequence logic is often, but not necessarily, coupled with the concept of a process state. Basically, the sequence logic determines when the process should proceed from the current state to the next and sometimes what the next state should be.

Sequence logic must encompass both normal and abnormal process operations. Thus, sequence logic is often viewed as consisting of two distinct but related parts:

1. *Normal logic.* This sequence logic provides for the normal or expected progression from one process state to another.
2. *Failure logic.* This logic provides for responding to abnormal conditions, such as equipment failures.

Of these, the failure logic can easily be the most demanding. The simplest approach is to stop or hold on any abnormal condition and let the process operator sort things out. However, this is not always acceptable. Some failures lead to hazardous conditions that require immediate action; waiting for the operator to decide what to do is not acceptable. The appropriate response to such situations is best determined in conjunction with the process hazards analysis.

No single approach has evolved as the preferred way to implement sequence logic. The approaches utilized include the following:

1. *Discrete logic.* Although sequence logic is different from discrete logic, sequence logic can be implemented using discrete logic capabilities. Simple sequences are commonly implemented as ladder diagrams in programmable logic controllers (PLCs). Sequence logic can also be implemented using the boolean logic functions provided by a distributed control system (DCS), although this approach is now infrequently pursued.
2. *Programming languages.* Traditional procedural languages do not provide the necessary constructs for implementing sequence logic. This necessitates one of the following:

a. Special languages. The necessary extensions for sequence logic are provided by extending the syntax of the programming language. This is the most common approach within distributed control systems. The early implementations used BASIC as the starting point for the extensions; the later implementations used C as the starting point. A major problem with this approach is portability, especially from one manufacturer to the next but sometimes from one product version to the next within the same manufacturer's product line.

b. Subroutine or function libraries. The facilities for sequence logic are provided via subroutines or functions that can be referenced from programs written in FORTRAN or C. This requires a general-purpose program development environment and excellent facilities to trap the inevitable errors in such programs. Operating systems with such capabilities have long been available on the larger computers, but not for the microprocessors utilized within DCSs. However, such operating systems are becoming more common within DCSs.

3. *State machines.* This technology is commonly applied within the discrete manufacturing industries. However, its migration to process batch applications has been limited.
4. *Graphical implementations.* For sequence logic, the flowchart traditionally used to represent the logic of computer programs must be extended to provide parallel execution paths. Such extensions have been implemented in a graphical representation generally referred to as a sequential function chart, which is a derivative of an earlier technology known as Grafcet. As process engineers have demonstrated a strong dislike for ladder logic, most PLC manufacturers now provide sequential function charts either in addition to or as an alternative to ladder logic. Many DCS manufacturers also provide sequential function charts either in addition to or as an alternative to special sequence languages.

INDUSTRIAL APPLICATIONS

An industrial example requiring simple sequence logic is the effluent tank with two sump pumps illustrated in Fig. 8-59. There are two sump pumps, A and B. The tank is equipped with three level switches, one for low level (LL), one for high level (LH), and one for high-high level (LHH). All level switches actuate on rising level. The logic is to be as follows:

1. When level switch LH actuates, start one sump pump. This must alternate between the sump pumps. If pump A is started on this occasion, then pump B must be started on the next occasion.

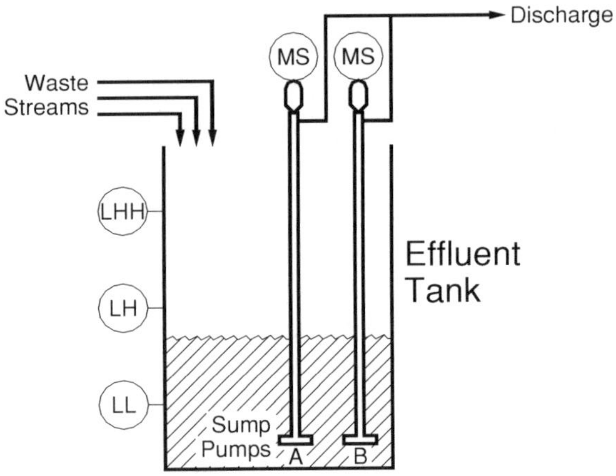

FIG. 8-59 Effluent tank process.

2. When level switch LHH actuates, start the other sump pump.
3. When level switch LL deactuates, stop all sump pumps.

Once a sump pump is started, it is not stopped until level switch LL deactuates. With this logic, one, both, or no sump pump may be running when the level is between LL and LH. Either one or both sump pumps may be running when the level is between LH and LHH.

Figure 8-60*a* presents the ladder logic implementation of the sequence logic. Ladder diagrams were originally developed for representing hardwired logic, but are now widely used in PLCs. The vertical bar on the left provides the source of power; the vertical bar on the right is ground. If a coil is connected between the power source and ground, the coil will be energized. If a circuit consisting of a set of contacts is inserted between the power source and the coil, the coil will be energized only if power can flow through the circuit. This will depend on the configuration of the circuit and the states of the contacts within the circuit. Ladder diagrams are constructed as rungs, with each rung consisting of a circuit of contacts and an output coil.

Contacts are represented as vertical bars. A vertical bar represents a normally open contact; power flows through this contact only if the device with which the contact is associated is actuated (energized). Vertical bars separated by a slash represent a normally closed contact; power flows through this contact only if the device with which the contact is associated is not actuated. The level switches actuate on rising level. If the vessel level is below the location of the switch, the normally open contact is open and the normally closed contact is closed. If the level is above the location of the switch, the normally closed contact is closed and the normally open contact is open.

The first rung in Fig. 8-60*a* is for pump A. It will run if one (or more) of the following conditions is true:

1. Level is above LH and pump A is the lead pump. A coil (designated as LeadIsB) will be subsequently provided to designate the pump to be started next (called the lead pump). If this coil is energized, pump B is the lead pump. Hence, pump A is to be started at LH if this coil is not energized, hence the use of the normally closed contact on coil LeadIsB in the rung of ladder logic for pump A.
2. Level is above LHH.
3. Pump A is running and the level is above LL.

The second rung is an almost identical circuit for pump B. The difference is the use of the normally open contact on the coil LeadIsB.

When implemented as hardwired logic, ladder diagrams are truly parallel logic; i.e., all circuits are active at all instants of time. But when ladder diagrams are implemented in PLCs, the behavior is slightly different. The ladder logic is scanned very rapidly (on the order of 100 times per second), which gives the appearance of parallel logic. But within a scan of ladder logic, the rungs are executed sequentially. This

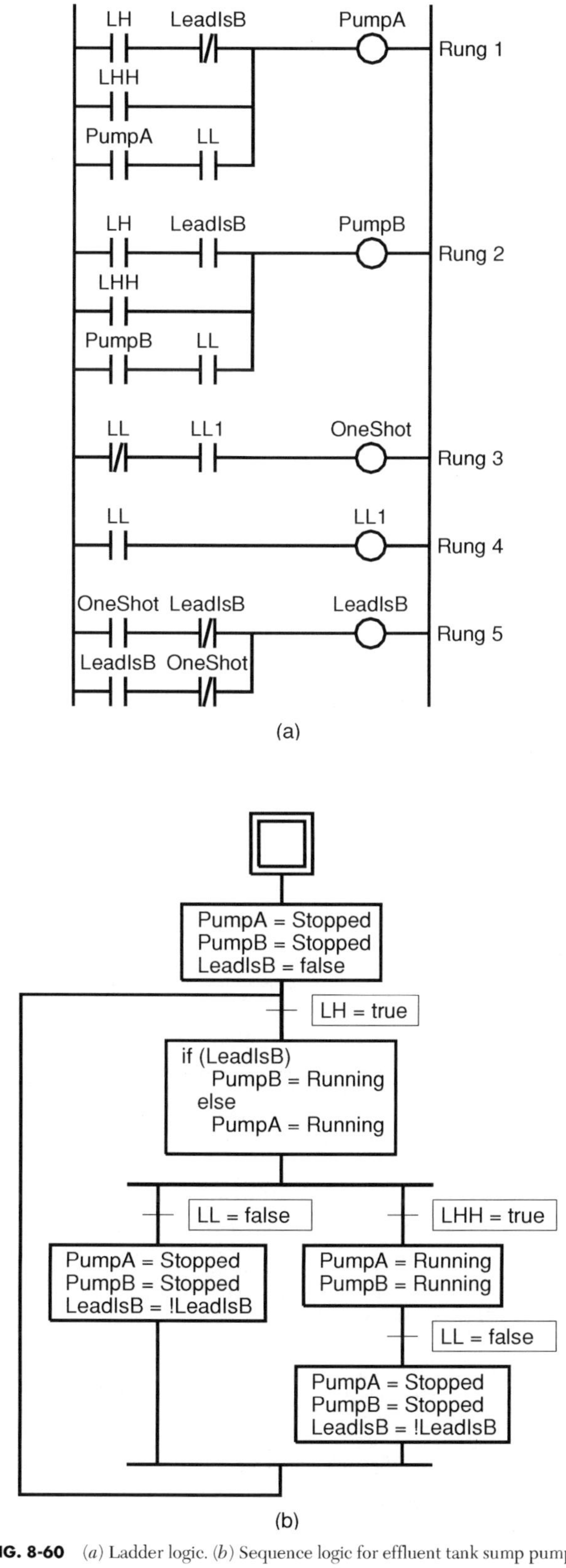

FIG. 8-60 (*a*) Ladder logic. (*b*) Sequence logic for effluent tank sump pumps.

permits constructs within ladder logic for PLCs that make no sense in hardwired circuits.

One such construct is for a "one-shot." Some PLCs provide this as a built-in function, but here it will be presented in terms of separate components. The one-shot is generated by the third rung of ladder logic in Fig. 8-60*a*. But first examine the fourth rung. The input LL drives the output coil LL1. This coil provides the state of level switch LL on the previous scan of ladder logic. This is used in the third rung to produce the one-shot. Output coil OneShot is energized if

1. LL is not actuated on this scan of ladder logic (note the use of the normally closed contact for LL)
2. LL was actuated on the previous scan of ladder logic (note the use of the normally open contact for LL1)

When LL deactuates, coil OneShot is energized for one scan of ladder logic. OneShot does not energize when LL actuates (a slight modification of the circuit would give a one-shot when LL actuates).

The one-shot is used in the fifth rung of ladder logic to toggle the lead pump. The output coil LeadIsB is energized provided that

1. LeadIsB is energized and OneShot is not energized. Once LeadIsB is energized, it remains energized until the next "firing" of the one-shot.
2. LeadIsB is not energized and OneShot is energized. This causes coil LeadIsB to change states each time the one-shot fires.

Ladder diagrams are ideally suited for representing discrete logic, such as required for interlocks. Sequence logic can be implemented via ladder logic, but usually with some tricks or gimmicks (the one-shot in Fig. 8-60*a* is such a gimmick). These are well known to those "skilled in the art" of PLC programming. But to others, they can be quite confusing.

Figure 8-60*b* provides a sequential function chart for the pumps. Sequential function charts consist of steps and transitions. A step consists of actions to be performed, as represented by statements. A transition consists of a logical expression. As long as the logical expression is false, the sequence logic remains at the transition. When the logical expression is true, the sequence logic proceeds to the step following the transition.

The basic constructs of sequential function charts are presented in Fig. 8-61. The basic construct of a sequential function chart is the step-transition-step. But also note the constructs for OR and AND. At the divergent OR, the logic proceeds on only one of the possible paths, specifically, the one whose transition is the first to attain the true condition. At the divergent AND, the logic proceeds on all paths simultaneously, and all must complete to proceed beyond the convergent AND. This enables sequential function charts to provide parallel logic.

In the sequential function chart in Fig. 8-60*b* for the pumps, the logic is initiated with both pumps stopped and pump A as the lead pump. When LH actuates, the lead pump is started. A divergent OR is used to create two paths:

1. If LL deactuates, both pumps are stopped and the lead pump is swapped.
2. If LHH actuates, both pumps are started (one is already running). Both remain running until LL deactuates, at which time both are stopped. The logic then loops to the transition for LH actuating.

Although not illustrated here, programming languages (either custom sequence languages or traditional languages extended by libraries of real-time functions) are a viable alternative for implementing the logic for the pumps. Graphical constructs such as ladder logic and sequential function charts are appealing to those uncomfortable with traditional programming languages. But in reality, these are programming methodologies.

BATCH REACTOR CONTROL

The reactors in flexible batch chemical plants usually present challenges. Many reactors have multiple mechanisms for heating and/or cooling. The reactor in Fig. 8-62 has three mechanisms:

1. Heat with steam.
2. Cool with cooling tower water.
3. Cool with chilled water.

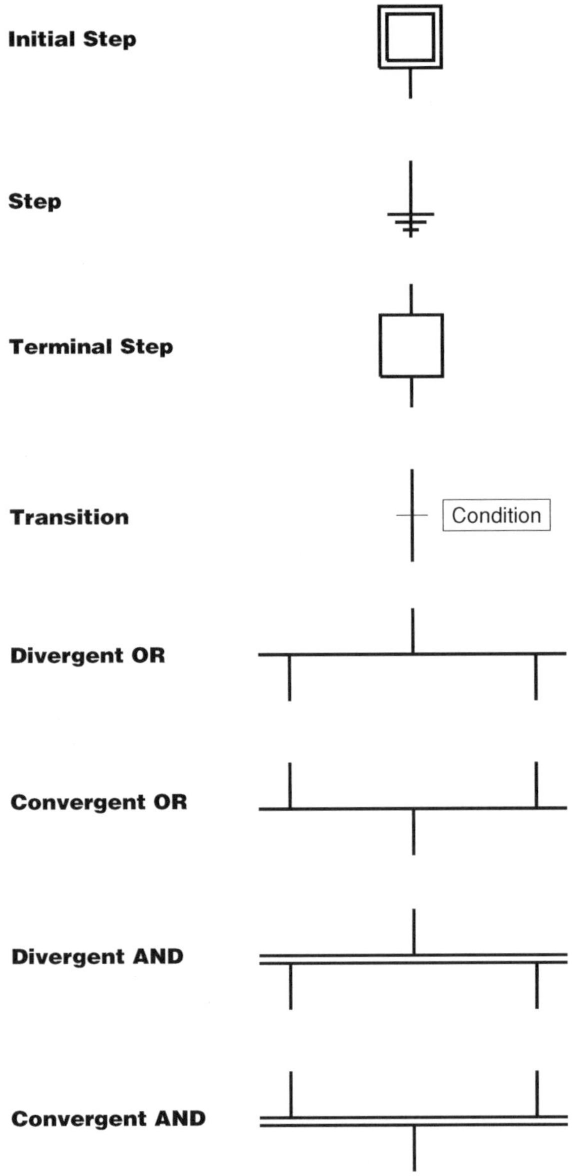

FIG. 8-61 Elements of sequential function charts.

Sometimes glycol is an option; occasionally liquid nitrogen is used to achieve even lower temperatures. Some jacket configurations require sequences for switching between the various modes for heating and cooling (the jacket has to be drained or depressurized before another medium can be admitted).

The reactor in Fig. 8-62 has three mechanisms for pressure control:

1. Vacuum
2. Atmospheric (using the vent and inert valves)
3. Pressure

Some reactors are equipped with multiple vacuum pumps, with different operating modes for moderate vacuum versus high vacuum. Sequence logic is usually required to start vacuum pumps and establish vacuum.

With three options for heating/cooling and three options for pressure, the reactor in Fig. 8-62 has nine combinations of operating modes.

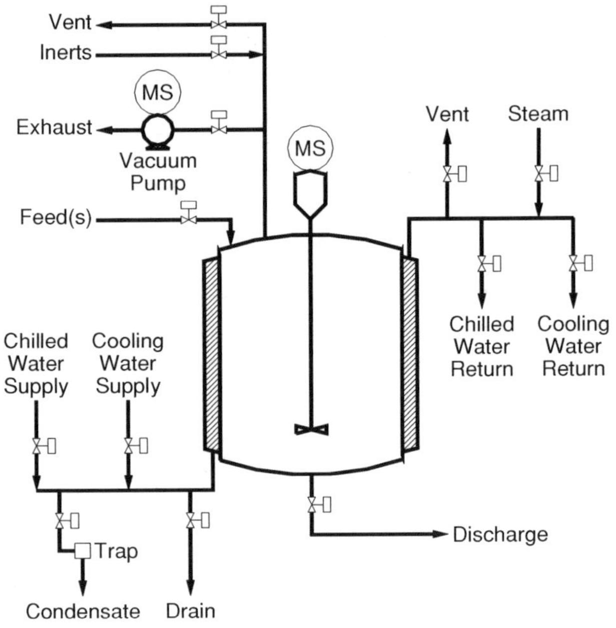

FIG. 8-62 Chemical reactor schematic.

In practice, this is actually a low number. This number increases with features such as

1. Recirculations or pump-arounds containing a heater and/or cooler
2. Reflux condensers that can be operated at total reflux (providing only cooling) or such that some component is removed from the reacting system

These further increase the number of possible combinations. Some combinations may not make sense, may not be used in current production operations, or otherwise can be eliminated. However, the net number of combinations that must be supported tends to be large.

The order in which the systems are activated usually depends on the product being manufactured. Sometimes heating/cooling and pressure control are established simultaneously, sometimes heating/cooling is established first and then pressure control, and sometimes pressure control is established first and then heating/cooling. One has to be very careful when imposing restrictions. Suppose no current products establish pressure control first and then establish heating/cooling. But what about the next product to be introduced? After all, this is a flexible batch facility.

Such challenging applications for recipe management and sequence logic require a detailed analysis of the production equipment and the operations conducted within that production equipment. While applications such as the sump pumps in the effluent tank can be pursued without it, a structured approach is essential in flexible batch facilities.

BATCH PRODUCTION FACILITIES

Especially for flexible batch applications, the batch logic must be properly structured in order to be implemented and maintained in a reasonable manner. An underlying requirement is that the batch process equipment be properly structured. The following structure is appropriate for most batch production facilities.

Plant A plant is the collection of production facilities at a geographic site. The production facilities at a site normally share warehousing, utilities, and the like.

Equipment Suite An equipment suite is the collection of equipment available for producing a group of products. Normally, this group of products is similar in certain respects. For example, they might all be manufactured from the same major raw materials. Within the equipment suite, material transfer and metering capabilities are available for these raw materials. The equipment suite contains all the necessary types of processing equipment (reactors, separators, and so on) required to convert the raw materials to salable products. A plant may consist of only one suite of equipment, but large plants usually contain multiple equipment suites.

Process Unit or Batch Unit A process unit is a collection of processing equipment that can, at least at certain times, be operated in a manner completely independent of the remainder of the plant. A process unit normally provides a specific function in the production of a batch of product. For example, a process unit might be a reactor complete with all associated equipment (jacket, recirculation pump, reflux condenser, and so on). However, each feed preparation tank is usually a separate process unit. With this separation, preparation of the feed for the next batch can be started as soon as the feed tank is emptied for the current batch.

All but the very simplest equipment suites contain multiple process units. The minimum number of process units is one for each type of processing equipment required to make a batch of product. However, many equipment suites contain multiple process units of each type. In such equipment suites, multiple batches and multiple production runs can be in progress at a given time.

Item of Equipment An item of equipment is a hardware item that performs a specific purpose. Examples are pumps, heat exchangers, agitators, and the like. A process unit could consist of a single item of equipment, but most process units consist of several items of equipment that must be operated in harmony to achieve the function expected of the process unit.

Device A device is the smallest element of interest to batch logic. Examples of devices include measurement devices and actuators.

STRUCTURED BATCH LOGIC

Flexible batch applications must be pursued by using a structured approach to batch logic. In such applications, the same processing equipment is used to make a variety of products. In most facilities, little or no proprietary technology is associated with the equipment itself; the proprietary technology is how this equipment is used to produce each of the products.

The primary objective of the structured approach is to separate cleanly the following two aspects of the batch logic:

Product Technology Basically, this encompasses the product technology, such as how to mix certain molecules to make other molecules. This technology ultimately determines the chemical and physical properties of the final product. The product recipe is the principal source for the product technology.

Process Technology The process equipment permits certain processing operations (e.g., heat to a specified temperature) to be undertaken. Each processing operation will involve certain actions (e.g., opening appropriate valves).

The need to keep these two aspects separated is best illustrated by a situation where the same product is to be made at different plants. While it is possible that the processing equipment at the two plants is identical, this is rarely the case. Suppose one plant uses steam for heating its vessels, but the other uses a hot oil system as the source of heat. When a product recipe requires that material be heated to a specified temperature, each plant can accomplish this objective, but each will go about it in quite different ways. The ideal case for a product recipe is as follows:

1. It contains all the product technology required to make a product.
2. It contains no equipment-dependent information, i.e., no process technology.

In the previous example, such a recipe would simply state that the product must be heated to a specified temperature. Whether heating is undertaken with steam or hot oil is irrelevant to the product technology. By restricting the product recipe to a given product technology, the same product recipe can be used to make products at different sites. At a given site, the specific approach to be used to heat a vessel is important. The traditional approach is for an engineer at each site to expand the product recipe into a document that explains in detail how the product is to be made at the specific site. This document goes by various names, although standard operating procedure or SOP is a common one. Depending on the level of detail to which it is written, the SOP could specify exactly which valves must be opened to heat the contents of a vessel. Thus, the SOP is site-dependent and contains both product technology and process technology.

In structuring the logic for a flexible batch application, the following organization permits product technology to be cleanly separated from process technology:

- A recipe consists of a formula and one or more processing operations. Ideally, only product technology is contained in a recipe.
- A processing operation consists of one or more phases. Ideally, only product technology is contained in a processing operation.
- A phase consists of one or more actions. Ideally, only process technology is contained in a phase.

In this structure, the recipe and processing operations would be the same at each site that manufactures the product. However, the logic that comprises each phase would be specific to a given site. In the heating example from above, each site would require a phase to heat the contents of the vessel. However, the logic within the phase at one site would accomplish the heating by opening the appropriate steam valves, while the logic at the other site would accomplish the heating by opening the appropriate hot oil valves.

Usually the critical part of structuring batch logic is the definition of the phases. There are two ways to approach this:

1. Examine the recipes for the current products for commonality, and structure the phases to reflect this commonality.
2 Examine the processing equipment to determine what processing capabilities are possible, and write phases to accomplish each possible processing capability.

There is the additional philosophical issue of whether to have a large number of simple phases with few options each, or a small number of complex phases with numerous options. The issues are analogous to structuring a complex computer program into subprograms. Each possible alternative has advantages and disadvantages.

As the phase contains no product technology, the implementation of a phase must be undertaken by those familiar with the process equipment. Furthermore, they should undertake this on the basis that the result will be used to make a variety of products, not just those that are initially contemplated. The development of the phase logic must also encompass all equipment-related safety issues. The phase should accomplish a clearly defined objective, so the implementers should be able to thoroughly consider all relevant issues in accomplishing this objective. The phase logic is defined in detail, implemented in the control system, and then thoroughly tested. Except when the processing equipment is modified, future modifications to the phase should be infrequent. The result should be a very dependable module that can serve as a building block for batch logic.

Even for flexible batch applications, a comprehensive menu of phases should permit most new products to be implemented by using currently existing phases. By reusing existing phases, numerous advantages accrue:

1. The engineering effort to introduce a new recipe at a site is reduced.
2. The product is more likely to be on-spec the first time, thus avoiding the need to dispose of off-spec product.
3. The new product can be supplied to customers sooner, hopefully before competitors can supply the product.

There is also a distinct advantage in maintenance. When a problem with a phase is discovered and the phase logic is corrected, the correction is effectively implemented in all recipes that use the phase. If a change is implemented in the processing equipment, the affected phases must be modified accordingly and then thoroughly tested. These modifications are also effectively implemented in all recipes that use these phases.

PROCESS MEASUREMENTS

GENERAL REFERENCES: Baker, *Flow Measurement Handbook*, Cambridge University Press, New York, 2000. Connell, *Process Instrumentation Applications Manual*, McGraw-Hill, New York, 1966. Dakin, and Culshaw (eds), *Optical Fiber Sensors: Applications, Analysis, and Future Trends,* vol. IV, Artech House, Norwood, Mass., 1997. Dolenc, "Choose the Right Flow Meter," *Chem. Engr. Prog.*, **92**(1): 22, 1996. Johnson, *Process Control Instrumentation Technology*, 6th ed., Prentice-Hall, Upper Saddle River, N.J., 2000. Liptak (ed.), *Instrument Engineers Handbook*, 3d ed., vol. 1: *Process Measurement*, Chilton Books, Philadelphia, 2000. Nichols, *On-Line Process Analyzers*, Wiley, New York, 1988. Seborg, Edgar, and Mellichamp, *Process Dynamics and Control,* Wiley, New York, 2004. Soloman, *Sensors Handbook*, McGraw-Hill, New York, 1999. Spitzer, *Flow Measurement*, 2d ed., ISA, Research Triangle Park, N.C., 2001.

GENERAL CONSIDERATIONS

Process measurements encompass the application of the principles of metrology to the process in question. The objective is to obtain values for the current conditions within the process and to make this information available in a form usable by the control system, process operators, or management information systems. The term *measured variable* or *process variable* designates the process condition that is being determined.

Process measurements fall into two categories:

1. *Continuous measurements.* An example of a continuous measurement is a level measurement device that determines the liquid level in a tank (e.g., in meters).
2. *Discrete measurements.* An example of a discrete measurement is a level switch that indicates the presence or absence of liquid at the location at which the level switch is installed.

In continuous processes, most process control applications rely on continuous measurements. In batch processes, many of the process control applications utilize discrete as well as continuous measurements. In both types of processes, the safety interlocks and process interlocks rely largely on discrete measurements.

Continuous Measurements In most applications, continuous measurements provide more information than discrete measurements. Basically, discrete measurements involve a yes/no decision, whereas continuous measurements may entail considerable signal processing.

The components of a typical continuous measurement device are as follows:

1. *Sensor.* This component produces a signal that is related in a known manner to the process variable of interest. The sensors in use today are primarily of the electrical analog variety, and the signal is in the form of a voltage, a resistance, a capacitance, or some other directly measurable electrical quantity. Prior to the mid-1970s, instruments tended to use sensors whose signal was mechanical and thus compatible with pneumatic technology. Since that time, the fraction of sensors that are digital has grown considerably, often eliminating the need for analog-to-digital conversion.
2. *Signal processing.* The signal from a sensor is usually related in a nonlinear fashion to the process variable of interest. For the output of the measurement device to be linear with respect to the process variable of interest, linearization is required. Furthermore, the signal from the sensor might be affected by variables other than the process variable. In this case, additional variables must be sensed, and the signal from the sensor compensated to account for the other variables. For example, reference junction compensation is required for thermocouples (except when used for differential temperature measurements).
3. *Transmitter.* The measurement device output must be a signal that can be transmitted over some distance. Where electronic analog transmission is used, the low range on the transmitter output is 4 mA, and the upper range is 20 mA. Microprocessor-based transmitters (often referred to as smart transmitters) are usually capable of transmitting the measured variable digitally in engineering units.

Accuracy and Repeatability Definitions of terminology pertaining to process measurements can be obtained from standards available from the Instrumentation, Systems, and Automation Society (ISA) and from the Scientific Apparatus Makers Association [now Measurement, Control, and Automation Association (MCAA)], both of which are updated periodically. An appreciation of accuracy and repeatability is especially important. Some applications depend on the accuracy of the instrument, but other applications depend on repeatability. Excellent accuracy implies excellent repeatability; however, an instrument can have poor accuracy but excellent repeatability. In some applications, this is acceptable, as discussed below.

Range and Span A continuous measurement device is expected to provide credible values of the measured value between a lower range and an upper range. The difference between the upper range and the lower range is the span of the measurement device. The maximum value for the upper range and the minimum value for the lower range depend on the principles on which the measurement device is based and on the design chosen by the manufacturer of the measurement device. If the measured variable is greater than the upper range or less than the lower range, the measured variable is said to be out of range or the measurement device is said to be overranged.

Accuracy Accuracy refers to the difference between the measured value and the true value of the measured variable. Unfortunately, the true value is never known, so in practice accuracy refers to the difference between the measured value and an accepted standard value for the measured variable.

Accuracy can be expressed in four ways:

1. As an absolute difference in the units of the measured variable
2. As a percent of the current reading
3. As a percent of the span of the measured variable
4. As a percent of the upper range of the span

For process measurements, accuracy as a percent of span is the most common.

Manufacturers of measurement devices always state the accuracy of the instrument. However, these statements always specify specific or reference conditions at which the measurement device will perform with the stated accuracy, with temperature and pressure most often appearing in the reference conditions. When the measurement device is applied at other conditions, the accuracy is affected. Manufacturers usually also provide some statements on how accuracy is affected when the conditions of use deviate from the referenced conditions in the statement of accuracy. Although appropriate calibration procedures can minimize some of these effects, rarely can they be totally eliminated. It is easily possible for such effects to cause a measurement device with a stated accuracy of 0.25 percent of span at reference conditions to ultimately provide measured values with accuracies of 1 percent or less. Microprocessor-based measurement devices usually provide better accuracy than do the traditional electronic measurement devices.

In practice, most attention is given to accuracy when the measured variable is the basis for billing, such as in custody transfer applications. However, whenever a measurement device provides data to any type of optimization strategy, accuracy is very important.

Repeatability Repeatability refers to the difference between the measurements when the process conditions are the same. This can also be viewed from the opposite perspective. If the measured values are the same, repeatability refers to the difference between the process conditions.

For regulatory control, repeatability is of major interest. The basic objective of regulatory control is to maintain uniform process operation. Suppose that on two different occasions, it is desired that the temperature in a vessel be 800°C. The regulatory control system takes appropriate actions to bring the measured variable to 800°C. The difference between the process conditions at these two times is determined by the repeatability of the measurement device.

In the use of temperature measurement for control of the separation in a distillation column, repeatability is crucial but accuracy is not. Composition control for the overhead product would be based on a measurement of the temperature on one of the trays in the rectifying section. A target would be provided for this temperature. However, at periodic intervals, a sample of the overhead product is analyzed in the laboratory and the information is provided to the process operator. Should this analysis be outside acceptable limits, the operator would adjust the set point for the temperature. This procedure effectively compensates for an inaccurate temperature measurement; however, the success of this approach requires good repeatability from the temperature measurement.

Dynamics of Process Measurements Especially where the measurement device is incorporated into a closed-loop control configuration, dynamics are important. The dynamic characteristics depend on the nature of the measurement device, and on the nature of components associated with the measurement device (e.g., thermowells and sample conditioning equipment). The term *measurement system* designates the measurement device and its associated components.

The following dynamics are commonly exhibited by measurement systems:

- *Time constants.* Where there is a capacity and a throughput, the measurement device will exhibit a time constant. For example, any temperature measurement device has a thermal capacity (mass times heat capacity) and a heat flow term (heat-transfer coefficient and area). Both the temperature measurement device and its associated thermowell will exhibit behavior typical of time constants.
- *Dead time.* Probably the best example of a measurement device that exhibits pure dead time (time delay) is the chromatograph, because the analysis is not available for some time after a sample is injected. Additional dead time results from the transportation lag within the sample system. Even continuous analyzer installations can exhibit dead time from the sample system.
- *Underdamped.* Measurement devices with mechanical components often have a natural harmonic and can exhibit underdamped behavior. The displacer type of level measurement device is capable of such behavior.

While the manufacturers of measurement devices can supply some information on the dynamic characteristics of their devices, interpretation is often difficult. Measurement device dynamics are quoted on varying bases, such as rise time, time to 63 percent response, settling time, etc. Even where the time to 63 percent response is quoted, it might not be safe to assume that the measurement device exhibits first-order behavior.

Where the manufacturer of the measurement device does not supply the associated equipment (thermowells, sample conditioning equipment, etc.), the user must incorporate the characteristics of these components to obtain the dynamics of the measurement system. An additional complication is that most dynamic data are stated for configurations involving reference materials such as water and air. The nature of the process material will affect the dynamic characteristics. For example, a thermowell will exhibit different characteristics when immersed in a viscous organic emulsion than when immersed in water. It is often difficult to extrapolate the available data to process conditions of interest.

Similarly, it is often impossible, or at least very difficult, to experimentally determine the characteristics of a measurement system under the conditions where it is used. It is certainly possible to fill an emulsion polymerization reactor with water and determine the dynamic characteristics of the temperature measurement system. However, it is not possible to determine these characteristics when the reactor is filled with the emulsion under polymerization conditions.

The primary impact of unfavorable measurement dynamics is on the performance of closed-loop control systems. This explains why most control engineers are very concerned with minimizing measurement dynamics, even though the factors considered in dynamics are often subjective.

Selection Criteria The selection of a measurement device entails a number of considerations given below, some of which are almost entirely subjective.

1. *Measurement span.* The measurement span required for the measured variable must lie entirely within the instrument's envelope of performance.

2. *Performance.* Depending on the application, accuracy, repeatability, or perhaps some other measure of performance is appropriate. Where closed-loop control is contemplated, speed of response must be included.

3. *Reliability.* Data available from the manufacturers can be expressed in various ways and at various reference conditions. Often, previous experience with the measurement device within the purchaser's organization is weighted most heavily.

4. *Materials of construction.* The instrument must withstand the process conditions to which it is exposed. This encompasses considerations such as operating temperatures, operating pressures, corrosion, and abrasion. For some applications, seals or purges may be necessary.

5. *Prior use.* For the first installation of a specific measurement device at a site, training of maintenance personnel and purchases of spare parts might be necessary.

6. *Potential for releasing process materials to the environment.* Fugitive emissions are receiving ever-increasing attention. Exposure considerations, both immediate and long-term, for maintenance personnel are especially important when the process fluid is either corrosive or toxic.

7. *Electrical classification.* Article 500 of the National Electric Code provides for the classification of the hazardous nature of the process area in which the measurement device will be installed. If the measurement device is not inherently compatible with this classification, suitable enclosures must be purchased and included in the installation costs.

8. *Physical access.* Subsequent to installation, maintenance personnel must have physical access to the measurement device for maintenance and calibration. If additional structural facilities are required, they must be included in the installation costs.

9. *Invasive or noninvasive.* The insertion of a probe can result in fouling problems and a need for maintenance. Probe location must be selected carefully for good accuracy and minimal fouling.

10. *Cost.* There are two aspects of the cost:

a. Initial purchase and installation (capital cost).

b. Recurring costs (operational expense). This encompasses instrument maintenance, instrument calibration, consumables (e.g., titrating solutions must be purchased for automatic titrators), and any other costs entailed in keeping the measurement device in service.

Calibration Calibration entails the adjustment of a measurement device so that the value from the measurement device agrees with the value from a standard. The International Standards Organization (ISO) has developed a number of standards specifically directed to calibration of measurement devices. Furthermore, compliance with the ISO 9000 standards requires that the working standard used to calibrate a measurement device be traceable to an internationally recognized standard such as those maintained by the National Institute of Standards and Technology (NIST).

Within most companies, the responsibility for calibrating measurement devices is delegated to a specific department. Often, this department may also be responsible for maintaining the measurement device. The specific calibration procedures depend on the type of measurement device. The frequency of calibration is normally predetermined, but earlier action may be dictated if the values from the measurement device become suspect.

Calibration of some measurement devices involves comparing the measured value with the value from the working standard. Pressure and differential pressure transmitters are calibrated in this manner. Calibration of analyzers normally involves using the measurement device to analyze a specially prepared sample whose composition is known. These and similar approaches can be applied to most measurement devices.

Flow is an important measurement whose calibration presents some challenges. When a flow measurement device is used in applications such as custody transfer, provision is made to pass a known flow through the meter. However, such a provision is costly and is not available for most in-process flowmeters. Without such a provision, a true calibration of the flow element itself is not possible. For orifice meters, calibration of the flowmeter normally involves calibration of the differential pressure transmitter, and the orifice plate is usually only inspected for deformation, abrasion, etc. Similarly, calibration of a magnetic flowmeter normally involves calibration of the voltage

TABLE 8-8 Online Measurement Options for Process Control

Temperature	Flow	Pressure	Level	Composition
Thermocouple	Orifice	Liquid column	Float-activated	Gas-liquid chromatography (GLC)
Resistance temperature detector (RTD)	Venturi	Elastic element	Chain gauge	Mass spectrometry (MS)
Filled-system thermometer	Rotameter	Bourdon tube	Lever	Magnetic resonance analysis (MRA)
Bimetal thermometer	Turbine	Bellow	Magnetically coupled	Infrared (IR) spectroscopy
Pyrometer	Vortex-shedding	Diaphragm	Head devices	Raman spectroscopy
Total radiation	Ultrasonic	Strain gauges	Bubble tube	Ultraviolet (uv) spectroscopy
Photoelectric	Magnetic	Piezoresistive transducers	Electrical (conductivity)	Thermal conductivity
Ratio	Thermal mass	Piezoelectric transducers	Sonic	Refractive index (RI)
Laser	Coriolis	Optical fiber	Laser	Capacitance probe
Surface acoustic wave	Target		Radiation	Surface acoustic wave
Semiconductor			Radar	Electrophoresis
				Electrochemical
				Paramagnetic
				Chemi/bioluminescence
				Tunable diode laser absorption

measurement circuitry, which is analogous to calibration of the differential pressure transmitter for an orifice meter.

In the next section we cover the major types of measurement devices used in the process industries, principally the "big five" measurements: temperature, flow rate, pressure, level, and composition, along with online physical property measurement techniques. Table 8-8 summarizes the different options under each of the principal measurements.

TEMPERATURE MEASUREMENTS

Measurement of the hotness or coldness of a body or fluid is commonplace in the process industries. Temperature-measuring devices utilize systems with properties that vary with temperature in a simple, reproducible manner and thus can be calibrated against known references (sometimes called secondary thermometers). The three dominant measurement devices used in automatic control are thermocouples, resistance thermometers, and pyrometers, and they are applicable over different temperature regimes.

Thermocouples Temperature measurements using thermocouples are based on the discovery by Seebeck in 1821 that an electric current flows in a continuous circuit of two different metallic wires if the two junctions are at different temperatures. The thermocouple may be represented diagrammatically as shown in Fig. 8-63. There A and B are the two metals, and T_1 and T_2 are the temperatures of the junctions. Let T_1 and T_2 be the reference junction (cold junction) and the measuring junction, respectively. If the thermoelectric current i flows in the direction indicated in Fig. 8-63, metal A is customarily referred to as thermoelectrically positive to metal B. Metal pairs used for thermocouples include platinum-rhodium (the most popular and accurate), chromel-alumel, copper-constantan, and iron-constantan. The thermal emf is a measure of the difference in temperature between T_2 and T_1. In control systems the reference junction is usually located at the emf-measuring device. The reference junction may be held at constant temperature such as in an ice bath or a thermostated oven, or it may be at ambient temperature but electrically compensated (cold junction compensated circuit) so that it appears to be held at a constant temperature.

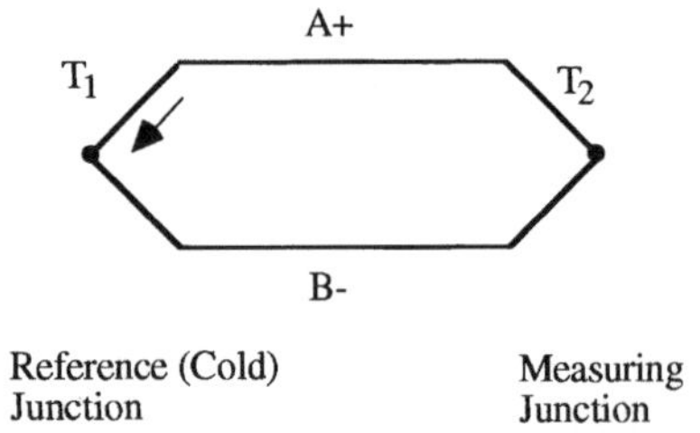

FIG. 8-63 Basic circuit of Seebeck effect.

Resistance Thermometers The resistance thermometer depends upon the inherent characteristics of materials to change in electrical resistance when they undergo a change in temperature. Industrial resistance thermometers are usually constructed of platinum, copper, or nickel, and more recently semiconducting materials such as thermistors are being used. Basically, a resistance thermometer is an instrument for measuring electrical resistance that is calibrated in units of temperature instead of in units of resistance (typically ohms). Several common forms of bridge circuits are employed in industrial resistance thermometry, the most common being the Wheatstone bridge. A resistance thermometer detector (RTD) consists of a resistance conductor (metal) which generally shows an increase in resistance with temperature. The following equation represents the variation of resistance with temperature (°C):

$$R_T = R_0(1 + a_1 T + a_2 T^2 + \cdots + a_n T^n) \tag{8-92}$$

R_0 = resistance at 0°C

The temperature coefficient of resistance, α_T is expressed as

$$\alpha_T = \frac{1}{R_T}\frac{dR_T}{dT} \tag{8-93}$$

For most metals α_T is positive. For many pure metals, the coefficient is essentially constant and stable over large portions of their useful range. Typical resistance versus temperature curves for platinum, copper, and nickel are given in Fig. 8-64, with platinum usually the metal of choice. Platinum has a useful range of −200 to 800°C, while nickel (−80 to 320°C) and copper (−100 to 100°C) are more limited. Detailed resistance versus temperature tables are available from the National Institute of Standards and Technology (NIST) and suppliers of resistance thermometers. Table 8-9 gives recommended temperature measurement ranges for thermocouples and RTDs. Resistance thermometers are receiving increased usage because they are about 10 times more accurate than thermocouples.

Thermistors Thermistors are nonlinear temperature-dependent resistors, and normally only the materials with negative temperature coefficient of resistance (NTC type) are used. The resistance is related to temperature as

$$R_T = R_{T_r}\exp\left[\beta\left(\frac{1}{T} - \frac{1}{T_r}\right)\right] \tag{8-94}$$

where α_T is a reference temperature, which is generally 298 K. Thus

$$\alpha_T = \frac{1}{R_T}\frac{dR_T}{dT} \tag{8-95}$$

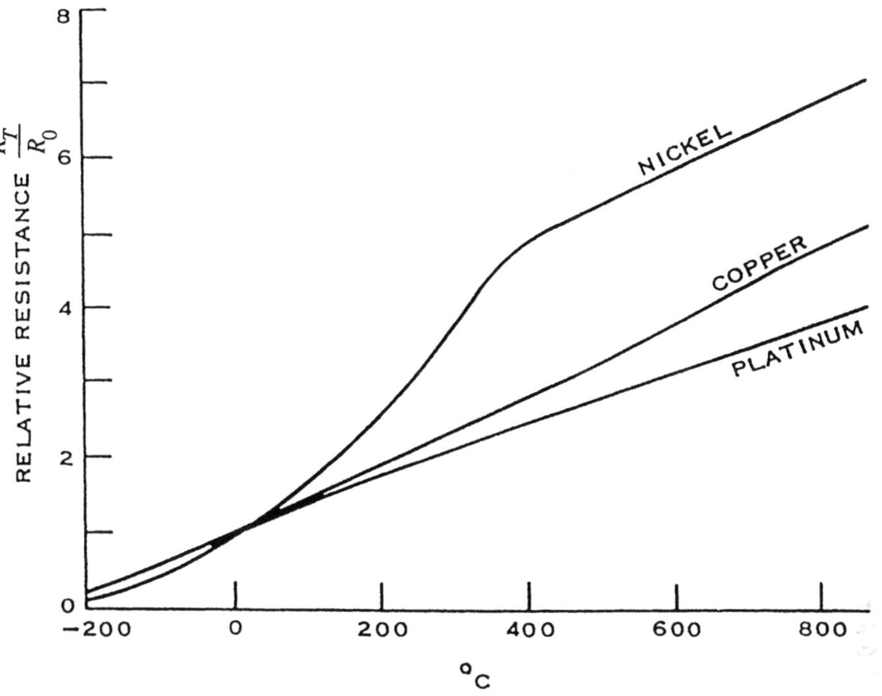

FIG. 8-64 Typical resistance thermometer curves for platinum, copper, and nickel wire, where R_T = resistance at temperature T and R_0 = resistance at 0°C.

The value of β is on the order of 4000, so at room temperature (298 K), α_T = −0.045 for thermistor and 0.0035 for 100-Ω platinum RTD. Compared with RTDs, NTC-type thermistors are advantageous in that the detector dimension can be made small, resistance value is higher (less affected by the resistances of the connecting leads), and it has higher temperature sensitivity and low thermal inertia of the sensor. Disadvantages of thermistors to RTDs include nonlinear characteristics and low measuring temperature range.

Filled-System Thermometers The filled-system thermometer is designed to provide an indication of temperature some distance removed from the point of measurement. The measuring element (bulb) contains a gas or liquid that changes in volume, pressure, or vapor pressure with temperature. This change is communicated through a capillary tube to a Bourdon tube or other pressure- or volume-sensitive device. The Bourdon tube responds so as to provide a motion related to the bulb temperature. Those systems that respond to volume changes are completely filled with a liquid. Systems that respond to pressure changes either are filled with a gas or are partially filled with a volatile liquid. Changes in gas or vapor pressure with changes in bulb temperatures are carried through the capillary to the Bourdon. The latter bulbs are sometimes constructed so that the capillary is filled with a nonvolatile liquid.

Fluid-filled bulbs deliver enough power to drive controller mechanisms and even directly actuate control valves. These devices are characterized by large thermal capacity, which sometimes leads to slow response, particularly when they are enclosed in a thermal well for process measurements. Filled-system thermometers are used extensively in industrial processes for a number of reasons. The simplicity of these devices allows rugged construction, minimizing the possibility of failure with a low level of maintenance, and inexpensive overall design of control equipment. In case of system failure, the entire unit must be replaced or repaired.

As they are normally used in the process industries, the sensitivity and percentage of span accuracy of these thermometers are generally the equal of those of other temperature-measuring instruments. Sensitivity and absolute accuracy are not the equal of those of short-span electrical instruments used in connection with resistance-thermometer bulbs. Also the maximum temperature is somewhat limited.

Bimetal Thermometers Thermostatic bimetal can be defined as a composite material made up of strips of two or more metals fastened together. This composite, because of the different expansion rates of its components, tends to change curvature when subjected to a change in temperature. With one end of a straight strip fixed, the other end deflects in proportion to the temperature change, the square of the length, and inversely as the thickness, throughout the linear portion of the deflection characteristic curve. If a bimetallic strip is wound into a helix or a spiral and one end is fixed, the other end will rotate when heat is applied. For a thermometer with uniform scale divisions, a bimetal must be designed to have linear deflection over the desired temperature range. Bimetal thermometers are used at temperatures ranging from 580 down to −180°C and lower. However, at the low temperatures

TABLE 8-9 Recommended Temperature Measurement Ranges for RTDs and Thermocouples

Resistance Thermometer Detectors (RTDs)	
100V Pt	−200–+850°C
120V Ni	−80–+320°C
Thermocouples	
Type B	700–+1820°C
Type E	−175–+1000°C
Type J	−185–+1200°C
Type K	−175–+1372°C
Type N	0–+1300°C
Type R	125–+1768°C
Type S	150–+1768°C
Type T	−170–+400°C

the rate of deflection drops off quite rapidly. Bimetal thermometers do not have long-time stability at temperatures above 430°C.

Pyrometers Planck's distribution law gives the radiated energy flux $q_b(\lambda, T)d\lambda$ in the wavelength range λ to $\lambda + d\lambda$ from a black surface:

$$q_b(\lambda, T) = \frac{C_1}{\lambda^5} \frac{1}{e^{C_2/\lambda T} - 1} \tag{8-96}$$

where $C_1 = 3.7418 \times 10^{10} \mu W \cdot \mu m^4 \cdot cm^{-2}$ and $C_2 = 14{,}388\ \mu m \cdot K$.

If the target object is a black body and if the pyrometer has a detector which measures the specific wavelength signal from the object, the temperature of the object can be exactly estimated from Eq. (8-96). While it is possible to construct a physical body that closely approximates black body behavior, most real-world objects are not black bodies. The deviation from a black body can be described by the spectral emissivity

$$\varepsilon_T = \frac{q(T)}{q_b(T)} \tag{8-97}$$

where $q(\lambda, T)$ is the radiated energy flux from a real body in the wavelength range λ to $\lambda + d\lambda$ and $0 < \varepsilon_{\lambda,T} < 1$. Integrating Eq. (8-96) over all wavelengths gives the Stefan-Boltzmann equation

$$q_b(T) = \int_0^\infty q_b(\lambda, T)\, d\lambda = \sigma T^4 \tag{8-98}$$

where σ is the Stefan-Boltzmann constant. Similar to Eq. (8-93), the emissivity ε_T for the total radiation is

$$\varepsilon_T = \frac{q(T)}{q_b(T)} \tag{8-99}$$

where $q(T)$ is the radiated energy flux from a real body with emissivity ε_T.

Total Radiation Pyrometers In total radiation pyrometers, the thermal radiation is detected over a large range of wavelengths from the object at high temperature. The detector is normally a thermopile, which is built by connecting several thermocouples in series to increase the temperature measurement range. The pyrometer is calibrated for black bodies, so the indicated temperature T_p should be converted for non–black body temperature.

Photoelectric Pyrometers Photoelectric pyrometers belong to the class of band radiation pyrometers. The thermal inertia of thermal radiation detectors does not permit the measurement of rapidly changing temperatures. For example, the smallest time constant of a thermal detector is about 1 ms while the smallest time constant of a photoelectric detector can be about 1 or 2 s. Photoelectric pyrometers may use photoconductors, photodiodes, photovoltaic cells, or vacuum photocells. Photoconductors are built from glass plates with thin film coatings of 1-μm thickness, using PbS, CdS, PbSe, or PbTe. When the incident radiation has the same wavelength as the materials are able to absorb, the captured incident photons free photoelectrons, which form an electric current. Photodiodes in germanium or silicon are operated with a reverse bias voltage applied. Under the influence of the incident radiation, their conductivity as well as their reverse saturation current is proportional to the intensity of the radiation within the spectral response band from 0.4 to 1.7 μm for Ge and from 0.6 to 1.1 μm for Si. Because of the above characteristics, the operating range of a photoelectric pyrometer can be either spectral or in specific band. Photoelectric pyrometers can be applied for a specific choice of the wavelength.

Disappearing Filament Pyrometers Disappearing filament pyrometers can be classified as spectral pyrometers. The brightness of a lamp filament is changed by adjusting the lamp current until the filament disappears against the background of the target, at which point the temperature is measured. Because the detector is the human eye, it is difficult to calibrate for online measurements.

Ratio Pyrometers The ratio pyrometer is also called the two-color pyrometer. Two different wavelengths are utilized for detecting the radiated signal. If one uses Wien's law for small values of AT, the detected signals from spectral radiant energy flux emitted at wavelengths λ_1 and λ_2 with emissivities ε_{λ_1} and ε_{λ_2} are

$$S_{\lambda_1} = KC_1\varepsilon_{\lambda_1}\lambda_1^{-5} \exp\left(\frac{-C_2}{\lambda_1 T}\right) \tag{8-100}$$

$$S_{\lambda_2} = KC_1\varepsilon_{\lambda_2}\lambda_2^{-5} \exp\left(\frac{-C_2}{\lambda_2 T}\right) \tag{8-101}$$

The ratio of the signals S_{λ_1} and S_{λ_2} is

$$\frac{S_{\lambda_1}}{S_{\lambda_2}} = \frac{\varepsilon_{\lambda_1}}{\varepsilon_{\lambda_2}} \left(\frac{\lambda_2}{\lambda_1}\right)^5 \exp\left[\frac{C_2}{T}\left(\frac{1}{\lambda_2} - \frac{1}{\lambda_1}\right)\right] \tag{8-102}$$

Nonblack or nongray bodies are characterized by the wavelength dependence of their spectral emissivity. Let T_c be defined as the temperature of the body corresponding to the temperature of a black body. If the ratio of its radiant intensities at wavelengths λ_1 and λ_2 equals the ratio of the radiant intensities of the non–black body, whose temperature is to be measured at the same wavelength, then Wien's law gives

$$\frac{\varepsilon_{\lambda_1}\exp(-C_2/\lambda_1 T)}{\varepsilon_{\lambda_2}\exp(-C_2/\lambda_2 T)} = \frac{\exp(-C_2/\lambda_1 T_c)}{\exp(-C_2/\lambda_2 T_c)} \tag{8-103}$$

where T is the true temperature of the body. Rearranging Eq. (8-103) gives

$$T = \left[\frac{\ln \varepsilon_{\lambda_1}/\varepsilon_{\lambda_2}}{C_2(1/\lambda_1 - 1/\lambda_2)} + \frac{1}{T_c}\right]^{-1} \tag{8-104}$$

For black or gray bodies, Eq. (8-104) reduces to

$$\frac{S_{\lambda_1}}{S_{\lambda_2}} = \left(\frac{\lambda_2}{\lambda_1}\right)^5 \exp\left[\frac{C_2}{T}\left(\frac{1}{\lambda_2} - \frac{1}{\lambda_1}\right)\right] \tag{8-105}$$

Thus by measuring S_{λ_1} and S_{λ_2}, the temperature T can be estimated.

Accuracy of Pyrometers Most of the temperature estimation methods for pyrometers assume that the object is either a gray body or has known emissivity values. The emissivity of the non–black body depends on the internal state or the surface geometry of the objects. Also the medium through which the thermal radiation passes is not always transparent. These inherent uncertainties of the emissivity values make the accurate estimation of the temperature of the target objects difficult. Proper selection of the pyrometer and accurate emissivity values can provide a high level of accuracy.

PRESSURE MEASUREMENTS

Pressure, defined as force per unit area, is usually expressed in terms of familiar units of weight-force and area or the height of a column of liquid which produces a like pressure at its base. Process pressure-measuring devices may be divided into three groups: (1) those based on the measurement of the height of a liquid column, (2) those based on the measurement of the distortion of an elastic pressure chamber, and (3) electrical sensing devices.

Liquid-Column Methods Liquid-column pressure-measuring devices are those in which the pressure being measured is balanced against the pressure exerted by a column of liquid. If the density of the liquid is known, the height of the liquid column is a measure of the pressure. Most forms of liquid-column pressure-measuring devices are commonly called manometers. When the height of the liquid is observed visually, the liquid columns are contained in glass or other transparent tubes. The height of the liquid column may be measured in length units or calibrated in pressure units. Depending on the pressure range, water and mercury are the liquids most frequently used. Because the density of the liquid used varies with temperature, the temperature must be taken into account for accurate pressure measurements.

Elastic Element Methods Elastic element pressure-measuring devices are those in which the measured pressure deforms some elastic material (usually metallic) within its elastic limit, the magnitude of the deformation being approximately proportional to the applied pressure. These devices may be loosely classified into three types: Bourdon tube, bellows, and diaphragm.

Bourdon Tube Elements Probably the most frequently used process pressure-indicating device is the C-spring Bourdon tube pressure gauge. Gauges of this general type are available in a wide variety of pressure ranges and materials of construction. Materials are selected on the basis of pressure range, resistance to corrosion by the process materials, and effect of temperature on calibration. Gauges calibrated with pressure, vacuum, compound (combination pressure and vacuum), and suppressed-zero ranges are available.

Bellows Element The bellows element is an axially elastic cylinder with deep folds or convolutions. The bellows may be used unopposed, or it may be restrained by an opposing spring. The pressure to be measured may be applied either to the inside or to the space outside the bellows, with the other side exposed to atmospheric pressure. For measurement of absolute pressure either the inside or the space outside of the bellows can be evacuated and sealed. Differential pressures may be measured by applying the pressures to opposite sides of a single bellows or to two opposing bellows.

Diaphragm Elements Diaphragm elements may be classified into two principal types: those that utilize the elastic characteristics of the diaphragm and those that are opposed by a spring or other separate elastic element. The first type usually consists of one or more capsules, each composed of two diaphragms bonded together by soldering, brazing, or welding. The diaphragms are flat or corrugated circular metallic disks. Metals commonly used in diaphragm elements include brass, phosphor bronze, beryllium copper, and stainless steel. Ranges are available from fractions of an inch of water to over 800 in (200 kPa) gauge. The second type of diaphragm is used for containing the pressure and exerting a force on the opposing elastic element. The diaphragm is a flexible or slack diaphragm of rubber, leather, impregnated fabric, or plastic. Movement of the diaphragm is opposed by a spring which determines the deflection for a given pressure. This type of diaphragm is used for the measurement of extremely low pressure, vacuum, or differential pressure.

Electrical Methods Electrical methods for pressure measurement include strain gauges, piezoresistive transducers, and piezoelectric transducers.

Strain Gauges When a wire or other electrical conductor is stretched elastically, its length is increased and its diameter is decreased. Both of these dimensional changes result in an increase in the electrical resistance of the conductor. Devices utilizing resistance-wire grids for measuring small distortions in elastically stressed materials are commonly called strain gauges. Pressure-measuring elements utilizing strain gauges are available in a wide variety of forms. They usually consist of one of the elastic elements described earlier to which one or more strain gauges have been attached to measure the deformation. There are two basic strain gauge forms: bonded and unbonded. Bonded strain gauges are bonded directly to the surface of the elastic element whose strain is to be measured. The unbonded strain gauge transducer consists of a fixed frame and an armature that moves with respect to the frame in response to the measured pressure. The strain gauge wire filaments are stretched between the armature and frame. The strain gauges are usually connected electrically in a Wheatstone bridge configuration.

Strain gauge pressure transducers are manufactured in many forms for measuring gauge, absolute, and differential pressures and vacuum. Full-scale ranges from 25.4 mm of water to 10,134 MPa are available. Strain gauges bonded directly to a diaphragm pressure-sensitive element usually have an extremely fast response time and are suitable for high-frequency dynamic pressure measurements.

Piezoresistive Transducers A variation of the conventional strain gauge pressure transducer uses bonded single-crystal semiconductor wafers, usually silicon, whose resistance varies with strain or distortion. Transducer construction and electrical configurations are similar to those using conventional strain gauges. A permanent magnetic field is applied perpendicular to the resonating sensor. An alternating current causes the resonator to vibrate, and the resonant frequency is a function of the pressure (tension) of the resonator. The principal advantages of piezoresistive transducers are a much higher bridge voltage output and smaller size. Full-scale output voltages of 50 to 100 mV/V of excitation are typical. Some newer devices provide digital rather than analog output.

Piezoelectric Transducers Certain crystals produce a potential difference between their surfaces when stressed in appropriate directions. Piezoelectric pressure transducers generate a potential difference proportional to a pressure-generated stress. Because of the extremely high electrical impedance of piezoelectric crystals at low frequency, these transducers are usually not suitable for measurement of static process pressures.

FLOW MEASUREMENTS

Flow, defined as volume per unit of time at specified temperature and pressure conditions, is generally measured by positive displacement or rate meters. The term *positive displacement meter* applies to a device in which the flow is divided into isolated measured volumes when the number of fillings of these volumes is counted in some manner. The term *rate meter* applies to all types of flowmeters through which the material passes without being divided into isolated quantities. Movement of the material is usually sensed by a primary measuring element that activates a secondary device. The flow rate is then inferred from the response of the secondary device by means of known physical laws or from empirical relationships.

The principal classes of flow-measuring instruments used in the process industries are variable-head, variable-area, positive-displacement, and turbine instruments; mass flowmeters; vortex-shedding and ultrasonic flowmeters; magnetic flowmeters; and more recently, Coriolis mass flowmeters. Head meters are covered in detail in Sec. 5.

Orifice Meter The most widely used flowmeter involves placing a fixed-area flow restriction (an orifice) in the pipe carrying the fluid. This flow restriction causes a pressure drop which can be related to flow rate. The sharp-edge orifice is popular because of its simplicity, low cost, and the large amount of research data on its behavior. For the orifice meter, the flow rate Q_a for a liquid is given by

$$Q_a = \frac{C_d A_2}{\sqrt{1 - (A_2/A_1)^2}} \cdot \sqrt{\frac{2(p_1 - p_2)}{\rho}} \qquad \text{(8-106)}$$

where $p_1 - p_2$ is the pressure drop, ρ is the density, A_1 is the pipe cross-sectional area, A_2 is the orifice cross-sectional area, and C_d is the discharge coefficient. The discharge coefficient C_d varies with the Reynolds number at the orifice and can be calibrated with a single fluid, such as water (typically $C_d \approx 0.6$). If the orifice and pressure taps are constructed according to certain standard dimensions, quite accurate (about 0.4 to 0.8 percent error) values of C_d may be obtained. Also note that the standard calibration data assume no significant flow disturbances such as elbows and valves for a certain minimum distance upstream of the orifice. The presence of such disturbances close to the orifice can cause errors of as much as 15 percent. Accuracy in measurements limits the meter to a flow rate range of 3:1. The orifice has a relatively large permanent pressure loss that must be made up by the pumping machinery.

Venturi Meter The venturi tube operates on exactly the same principle as the orifice [see Eq. (8-102)]. Discharge coefficients of venturis are larger than those for orifices and vary from about 0.94 to 0.99. A venturi gives a definite improvement in power losses over an orifice and is often indicated for measuring very large flow rates, where power losses can become economically significant. The initial higher cost of a venturi over an orifice may thus be offset by reduced operating costs.

Rotameter A rotameter consists of a vertical tube with a tapered bore in which a float changes position with the flow rate through the tube. For a given flow rate, the float remains stationary because the vertical forces of differential pressure, gravity, viscosity, and buoyancy are balanced. The float position is the output of the meter and can be made essentially linear with flow rate by making the tube area vary linearly with the vertical distance.

Turbine Meter If a turbine wheel is placed in a pipe containing a flowing fluid, its rotary speed depends on the flow rate of the fluid. A turbine can be designed whose speed varies linearly with flow rate. The speed can be measured accurately by counting the rate at which turbine blades pass a given point, using magnetic pickup to produce voltage pulses. By feeding these pulses to an electronic pulse rate meter one can measure flow rate by summing the pulses during a timed interval. Turbine meters are available with full-scale flow rates ranging from about 0.1 to 30,000 gal/min for liquids and 0.1 to 15,000 ft^3/min for air. Nonlinearity can be less than 0.05 percent in the larger sizes. Pressure drop across the meter varies with the square of flow rate and is about 3 to 10 psi at full flow. Turbine meters can follow flow transients quite accurately since their fluid/mechanical time constant is on the order of 2 to 10 ms.

Vortex-Shedding Flowmeters These flowmeters take advantage of vortex shedding, which occurs when a fluid flows past a nonstreamlined object (a blunt body). The flow cannot follow the shape of the object and separates from it, forming turbulent vortices or eddies at the object's side surfaces. As the vortices move downstream, they grow in size and are eventually shed or detached from the object. Shedding takes place alternately at either side of the object, and the rate of vortex formation and shedding is directly proportional to the volumetric flow rate. The vortices are counted and used to develop a signal linearly proportional to the flow rate. The digital signals can easily be totaled over an interval of time to yield the flow rate. Accuracy can be maintained regardless of density, viscosity, temperature, or pressure when the Reynolds number is greater than 10,000. There is usually a low flow cutoff point below which the meter output is clamped at zero. This flowmeter is recommended for use with relatively clean, low-viscosity liquids, gases, and vapors, and rangeability of 10:1 to 20:1 is typical. A sufficient length of straight-run pipe is necessary to prevent distortion in the fluid velocity profile.

Ultrasonic Flowmeters An ultrasonic flowmeter is based upon the variable time delays of received sound waves which arise when a flowing liquid's rate of flow is varied. Two fundamental measurement techniques, depending upon liquid cleanliness, are generally used. In the first technique two opposing transducers are inserted in a pipe so that one transducer is downstream from the other. These transducers are then used to measure the difference between the velocity at which the sound travels with the direction of flow and the velocity at which it travels against the direction of flow. The differential velocity is measured either by (1) direct time delays using sound wave burst or (2) frequency shifts derived from beat-together, continuous signals. The frequency measurement technique is usually preferred because of its simplicity and independence of the liquid static velocity. A relatively clean liquid is required to preserve the uniqueness of the measurement path.

In the second technique, the flowing liquid must contain scatters in the form of particles or bubbles which will reflect the sound waves. These scatters should be traveling at the velocity of the liquid. A Doppler method is applied by transmitting sound waves along the flow path and measuring the frequency shift in the returned signal from the scatters in the process fluid. This frequency shift is proportional to liquid velocity.

Magnetic Flowmeters The principle behind these flowmeters is Faraday's law of electromagnetic inductance. The magnitude of the voltage induced in a conductive medium moving at right angles through a magnetic field is directly proportional to the product of the magnetic flux density, the velocity of the medium, and the path length between the probes. A minimum value of fluid conductivity is required to make this approach viable. The pressure of multiple phases or undissolved solids can affect the accuracy of the measurement if the velocities of the phases are different from that for straight-run pipe. Magmeters are very accurate over wide flow ranges and are especially accurate at low flow rates. Typical applications include metering viscous fluids, slurries, or highly corrosive chemicals. Because magmeters should be filled with fluid, the preferred installation is in vertical lines with flow going upward. However, magmeters can be used in tight piping schemes where it is impractical to have long pipe runs, typically requiring lengths equivalent to five or more pipe diameters.

Coriolis Mass Flowmeters Coriolis mass flowmeters utilize a vibrating tube in which Coriolis acceleration of a fluid in a flow loop can be created and measured. They can be used with virtually any liquid and are extremely insensitive to operating conditions, with pressure ranges over 100:1. These meters are more expensive than volumetric meters and range in size from $\frac{1}{16}$ to 6 in. Due to the circuitous path of flow through the meter, Coriolis flowmeters exhibit higher than average pressure changes. The meter should be installed so that it will remain full of fluid, with the best installation in a vertical pipe with flow going upward. There is no Reynolds number limitation with this meter, and it is quite insensitive to velocity profile distortions and swirl, hence there is no requirement for straight piping upstream.

Thermal Mass Flowmeters The trend in the chemical process industries is toward increased usage of mass flowmeters that are independent of changes in pressure, temperature, viscosity, and density. Thermal mass meters are widely used in semiconductor manufacturing and in bioprocessing for control of low flow rates (called mass flow controllers, or MFCs). MFCs measure the heat loss from a heated element, which varies with flow rate, with an accuracy of ± 1 percent. Capacitance probes measure the dielectric constant of the fluid and are useful for flow measurements of slurries and other two-phase flows.

LEVEL MEASUREMENTS

The measurement of level can be defined as the determination of the location of the interface between two fluids, separable by gravity, with respect to a fixed reference plane. The most common level measurement is that of the interface between a liquid and a gas. Other level measurements frequently encountered are the interface between two liquids, between a granular or fluidized solid and a gas, and between a liquid and its vapor.

A commonly used basis for classification of level devices is as follows: float-actuated, displacer, and head devices, and a miscellaneous group which depends mainly on fluid characteristics.

Float-Actuated Devices Float-actuated devices are characterized by a buoyant member which floats at the interface between two fluids. Because a significant force is usually required to move the indicating mechanism, float-actuated devices are generally limited to liquid-gas interfaces. By properly weighting the float, they can be used to measure liquid-liquid interfaces. Float-actuated devices may be classified on the basis of the method used to couple the float motion to the indicating system, as discussed below.

Chain or Tape Float Gauge In these types of gauges, the float is connected to the indicating mechanism by means of a flexible chain or tape. These gauges are commonly used in large atmospheric storage tanks. The gauge-board type is provided with a counterweight to keep the tape or chain taut. The tape is stored in the gauge head on a spring-loaded reel. The float is usually a pancake-shaped hollow metal float, with guide wires from top to bottom of the tank to constrain it.

Lever and Shaft Mechanisms In pressurized vessels, float-actuated lever and shaft mechanisms are frequently used for level measurement. This type of mechanism consists of a hollow metal float and lever attached to a rotary shaft which transmits the float motion to the outside of the vessel through a rotary seal.

Magnetically Coupled Devices A variety of float-actuated level devices which transmit the float motion by means of magnetic coupling have been developed. Typical of this class of devices are magnetically operated level switches and magnetic-bond float gauges. A typical magnetic-bond float gauge consists of a hollow magnet-carrying float which rides along a vertical nonmagnetic guide tube. The follower magnet is connected and drives an indicating dial similar to that on a conventional tape float gauge. The float and guide tube are in contact with the measured fluid and come in a variety of materials for resistance to corrosion and to withstand high pressures or vacuum. Weighted floats for liquid-liquid interfaces are available.

Head Devices A variety of devices utilize hydrostatic head as a measure of level. As in the case of displacer devices, accurate level measurement by hydrostatic head requires an accurate knowledge of the densities of both heavier-phase and lighter-phase fluids. The majority of this class of systems utilize standard pressure and differential pressure measuring devices.

Bubble Tube Systems The commonly used bubble tube system sharply reduces restrictions on the location of the measuring element. To eliminate or reduce variations in pressure drop due to the gas flow rate, a constant differential regulator is commonly employed to maintain a constant gas flow rate. Because the flow of gas through the bubble tube prevents entry of the process liquid into the measuring system, this technique is particularly useful with corrosive or viscous liquids, liquids subject to freezing, and liquids containing entrained solids.

Electrical Methods Two electrical characteristics of fluids, conductivity and dielectric constant, are frequently used to distinguish between two phases for level measurement purposes. An application of electrical conductivity is the fixed-point level detection of a conductive liquid such as high and low water levels. A voltage is applied between two electrodes inserted into the vessel at different levels. When both electrodes are immersed in the liquid, a current flows. Capacitance-type level measurements are based on the fact that the electrical capacitance between two electrodes varies with the dielectric constant of the material between them. A typical continuous level measurement system consists of a rod electrode positioned vertically in a vessel, the other electrode usually being the metallic vessel wall. The electrical capacitance between the electrodes is a measure of the height of the interface along the rod electrode. The rod is usually conductively insulated from process fluids by a coating of plastic. The dielectric constants of most liquids and solids are markedly higher than those of gases and vapors (by a factor of 2 to 5). The dielectric constant of water and other polar liquids is 10 to 20 times that of hydrocarbons and other nonpolar liquids.

Thermal Methods Level-measuring systems may be based on the difference in thermal characteristics between the fluids, such as temperature or thermal conductivity. A fixed-point level sensor based on the difference in thermal conductivity between two fluids consists of an electrically heated thermistor inserted into the vessel. The temperature of the thermistor and consequently its electrical resistance increase as the thermal conductivity of the fluid in which it is immersed decreases. Because the thermal conductivity of liquids is markedly higher than that of vapors, such a device can be used as a point level detector for liquid-vapor interface.

Sonic Methods A fixed-point level detector based on sonic propagation characteristics is available for detection of a liquid-vapor interface. This device uses a piezoelectric transmitter and receiver, separated by a short gap. When the gap is filled with liquid, ultrasonic energy is transmitted across the gap, and the receiver actuates a relay. With a vapor filling the gap, the transmission of ultrasonic energy is insufficient to actuate the receiver.

Laser Level Transmitters These are designed for bulk solids, slurries, and opaque liquids. A laser near the vessel top fires a short pulse of light down to the surface of the process liquid, where it reflects back to a detector at the vessel top. A timing circuit measures the elapsed time and calculates the fluid depth. Lasers are attractive because lasers have no false echoes and can be directed through tight spaces.

Radar Level Transmitters Radar systems operate by beaming microwaves downward, from either a horn or parabolic dish located on top of the vessel. The signal reflects off the fluid surface back to the source after it detects a change in dielectric constant from the vapor to the fluid. The round-trip time is proportional to the distance to the fluid level. Guided-wave radar systems provide a rigid probe or flexible cable to guide the microwave down the height of the tank and back. Guided-wave radar is much more efficient than open-air radar because the guide provides a more focused energy path.

PHYSICAL PROPERTY MEASUREMENTS

Physical property measurements are sometimes equivalent to composition analyzers, because the composition can frequently be inferred from the measurement of a selected physical property.

Density and Specific Gravity For binary or pseudobinary mixtures of liquids or gases or a solution of a solid or gas in a solvent, the density is a function of the composition at a given temperature and pressure. Specific gravity is the ratio of the density of a noncompressible substance to the density of water at the same physical conditions. For nonideal solutions, empirical calibration will give the relationship between density and composition. Several types of measuring devices are described below.

Liquid Column Density may be determined by measuring the gauge pressure at the base of a fixed-height liquid column open to the atmosphere. If the process system is closed, then a differential pressure measurement is made between the bottom of the fixed-height liquid column and the vapor over the column. If vapor space is not always present, the differential pressure measurement is made between the bottom and top of a fixed-height column with the top measurement being made at a point below the liquid surface.

Displacement There are a variety of density measurement devices based on displacement techniques. A hydrometer is a constant-weight, variable-immersion device. The degree of immersion, when the weight of the hydrometer equals the weight of the displaced liquid, is a measure of the density. The hydrometer is adaptable to manual or automatic usage. Another modification includes a magnetic float suspended below a solenoid, the varying magnetic field maintaining the float at a constant distance from the solenoid. Change in position of the float, resulting from a density change, excites an electrical system which increases or decreases the current through the solenoid.

Direct Mass Measurement One type of densitometer measures the natural vibration frequency and relates the amplitude to changes in density. The density sensor is a V-shaped tube held stationary at its node points and allowed to vibrate at its natural frequency. At the curved end of the V is an electrochemical device that periodically strikes the tube. At the other end of the V the fluid is continuously passed through the tube. Between strikes, the tube vibrates at its natural frequency. The frequency changes directly in proportion to changes in density. A pickup device at the curved end of the V measures the frequency and electronically determines the fluid density. This technique is useful because it is not affected by the optical properties of the fluid. However, particulate matter in the process fluid can affect the accuracy.

Radiation-Density Gauges Gamma radiation may be used to measure the density of material inside a pipe or process vessel. The equipment is basically the same as for level measurement, except that here the pipe or vessel must be filled over the effective, irradiated sample volume. The source is mounted on one side of the pipe or vessel and the detector on the other side with appropriate safety radiation shielding surrounding the installation. Cesium 137 is used as the radiation source for path lengths under 610 mm (24 in) and cobalt 60 above 610 mm. The detector is usually an ionization gauge. The absorption of the gamma radiation is a function of density. Since the absorption path includes the pipe or vessel walls, an empirical calibration is used. Appropriate corrections must be made for the source intensity decay with time.

Viscosity Continuous viscometers generally measure either the resistance to flow or the drag or torque produced by movement of an element (moving surface) through the fluid. Each installation is normally applied over a narrow range of viscosities. Empirical calibration over this range allows use on both newtonian and nonnewtonian fluids. One such device uses a piston inside a cylinder. The hydrodynamic pressure of the process fluid raises the piston to a preset height. Then the inlet valve closes, the piston is allowed to free-fall, and the time of travel (typically a few seconds) is a measure of viscosity. Other geometries include the rotation of a spindle inside a sample chamber and a vibrating probe immersed in the fluid. Because viscosity depends on temperature, the viscosity measurement must be thermostated with a heater or cooler.

Refractive Index When light travels from one medium (e.g., air or glass) into another (e.g., a liquid), it undergoes a change of velocity and, if the angle of incidence is not 90°, a change of direction. For a given interface, angle, temperature, and wavelength of light, the amount of deviation or refraction will depend on the composition of the liquid. If the sample is transparent, the normal method is to measure the refraction of light transmitted through the glass-sample interface. If the sample is opaque, the reflectance near the critical angle at a glass-sample interface is measured. In an online refractometer the process fluid is separated from the optics by a prism material. A beam of light is focused on a point in the fluid which creates a conic section of light at the prism, striking the fluid at different angles (greater than or less than the critical angle). The critical angle depends on the species concentrations; as the critical angle changes, the proportions of reflected and refracted light change. A photodetector produces a voltage signal proportional to the light refracted, when compared to a reference signal. Refractometers can be used with opaque fluids and in streams that contain particulates.

Dielectric Constant The dielectric constant of material represents its ability to reduce the electric force between two charges separated in space. This property is useful in process control for polymers, ceramic materials, and semiconductors. Dielectric constants are measured with respect to vacuum (1.0); typical values range from 2 (benzene) to 33 (methanol) to 80 (water). The value for water is higher than that for most plastics. A measuring cell is made of glass or some other insulating material and is usually doughnut-shaped, with the cylinders coated with metal, which constitute the plates of the capacitor.

Thermal Conductivity All gases and vapor have the ability to conduct heat from a heat source. At a given temperature and physical environment, radiation and convection heat losses will be stabilized, and the temperature of the heat source will mainly depend on the thermal conductivity and thus the composition of the surrounding gases. Thermal conductivity analyzers normally consist of a sample cell and a reference cell, each containing a combined heat source and detector. These cells are normally contained in a metal block with two small cavities in which the detectors are mounted. The sample flows through the sample cell cavity past the detector. The reference cell is an identical cavity with a detector through which a known gas flows. The combined heat source and detectors are normally either wire filaments or thermistors heated by a constant current. Because their resistance is a function of temperature, the sample detector resistance will vary with sample composition while the reference detector resistance will remain constant. The output from the detector bridge will be a function of sample composition.

CHEMICAL COMPOSITION ANALYZERS

Chemical composition is generally the most challenging online measurement. Before the era of online analyzers, messengers were required to deliver samples to the laboratory for analysis and to return the results to the control room. The long time delay involved prevented process adjustment from being made, affecting product quality. The development of online analyzers has automated this approach and reduced the analysis time. However, manual sampling is still frequently employed, especially in the specialty chemical industry where few instruments are commercially available. It is not unusual for a chemical composition analysis system to cost over $100,000, so it is important to assess the payback of such an investment versus the cost of manual sampling. Potential quality improvements can be an important consideration.

A number of composition analyzers used for process monitoring and control require chemical conversion of one or more sample components preceding quantitative measurement. These reactions include formation of suspended solids for turbidimetric measurement, formation of colored materials for colorimetric detection, selective oxidation or reduction for electrochemical measurement, and formation of electrolytes for measurement by electrical conductance. Some nonvolatile materials may be separated and measured by gas chromatography after conversion to volatile derivatives.

Chromatographic Analyzers These analyzers are widely used for the separation and measurement of volatile compounds and of compounds that can be quantitatively converted to volatile derivatives. The compounds to be measured are separated by placing a portion of the sample in a chromatographic column and carrying the compounds through the column with a gas stream, called gas chromatography, or GC. As a result of the different affinities of the sample components for the column packing, the compounds emerge successively as binary mixtures with the carrier gas. A detector at the column outlet measures a specific physical property that can be related to the concentrations of the compounds in the carrier gas. Both the concentration peak height and the peak height-time integral, i.e., peak area, can be related to the concentration of the compound in the original sample. The two detectors most commonly used for process chromatographs are the thermal conductivity detector and the hydrogen flame ionization detector. Thermal conductivity detectors, discussed earlier, require calibration for the thermal response of each compound. Hydrogen flame ionization detectors are more complicated than thermal conductivity detectors but are capable of 100 to 10,000 times greater sensitivity for hydrocarbons and organic compounds. For ultrasensitive detection of trace impurities, carrier gases must be specially purified.

Typically, all components can be analyzed in a 5- to 10-min time period (although miniaturized GCs are faster). High-performance liquid chromatography (HPLC) can be used to measure dissolved solute levels, including proteins.

Infrared Analyzers Many gaseous and liquid compounds absorb infrared radiation to some degree. The degree of absorption at specific wavelengths depends on molecular structure and concentration. There are two common detector types for nondispersive infrared analyzers. These analyzers normally have two beams of radiation, an analyzing and a reference beam. One type of detector consists of two gas-filled cells separated by a diaphragm. As the amount of infrared energy absorbed by the detector gas in one cell changes, the cell pressure changes. This causes movement in the diaphragm, which in turn causes a change in capacitance between the diaphragm and a reference electrode. This change in electrical capacitance is measured as the output. The second type of detector consists of two thermopiles or two bolometers, one in each of the two radiation beams. The infrared radiation absorbed by the detector is measured by a differential thermocouple output or a resistance thermometer (bolometer) bridge circuit.

There are two common detector types for nondispersive analyzers. These analyzers normally have two beams of radiation, an analyzing and a reference beam. One type of detector consists of two gas-filled cells separated by a diaphragm. As the amount of infrared energy absorbed by the detector gas in one cell changes, the cell pressure changes. This causes movement in the diaphragm, which in turn causes a change in capacitance between the diaphragm and a reference electrode. This change in electrical capacitance is measured as the output. The second type of detector consists of two thermopiles or two bolometers, one in each of the two radiation beams. The infrared radiation absorbed by the detector is measured by a differential thermocouple output or a resistance thermometer (bolometer) bridge circuit. With gas-filled detectors, a chopped light system is normally used in which one side of the detector sees the source through the analyzing beam and the other side sees through the reference beam, alternating at a frequency of a few hertz.

Ultraviolet and Visible-Radiation Analyzers Many gas and liquid compounds absorb radiation in the near-ultraviolet or visible region. For example, organic compounds containing aromatic and carbonyl structural groups are good absorbers in the ultraviolet region. Also many inorganic salts and gases absorb in the ultraviolet or visible region. In contrast, straight chain and saturated hydrocarbons, inert gases, air, and water vapor are essentially transparent. Process analyzers are designed to measure the absorbance in a particular wavelength band. The desired band is normally isolated by means of optical filters. When the absorbance is in the visible region, the term *colorimetry* is used. A phototube is the normal detector. Appropriate optical filters are used to limit the energy reaching the detector to the desired level and the desired wavelength region. Because absorption by the sample is logarithmic if a sufficiently narrow wavelength region is used, an exponential amplifier is sometimes used to compensate and produce a linear output.

Paramagnetism A few gases including O_2, NO, and NO_2 exhibit paramagnetic properties as a result of unpaired electrons. In a nonuniform magnetic field, paramagnetic gases, because of their magnetic susceptibility, tend to move toward the strongest part of the field, thus displacing diamagnetic gases. Paramagnetic susceptibility of these gases decreases with temperature. These effects permit measurement of the concentration of the strongest paramagnetic gas, oxygen. An oxygen analyzer uses a dumbbell suspended in the magnetic field which is repelled or attracted toward the magnetic field depending on the magnetic susceptibility of the gas.

Other Analyzers Mass spectroscopy (MS) determines the partial pressures of gases in a mixture of directing ionized gases into a detector under a vacuum (10^{-6} torr), and the gas phase composition is then monitored more or less continuously based on the molecular weight of the species (Nichols, 1988). Sometimes GC is combined with MS to obtain a higher level of discrimination of the components present. Fiber-optic sensors are attractive options (although higher-cost) for acquiring measurements in harsh environments such as high temperature or pressure. The transducing technique used by these sensors is optical and does not involve electric signals, so they are immune to electromagnetic interference. Raman spectroscopy uses fiber optics and involves pulsed light scattering by molecules. It has a wide variety of applications in process control. Workman, Koch, and Veltkamp, *Anal. Chem.*, **75:** 2859, 2003.

Significant advances have occurred during the past decade to miniaturize the size of the measurement system in order to make online analysis economically feasible and to reduce the time delays that often are present in analyzers. Recently, chemical sensors have been placed on microchips, even those requiring multiple physical, chemical, and biochemical steps (such as electrophoresis) in the analysis. This device has been called *lab-on-a-chip*. The measurements of chemical composition can be direct or indirect, the latter case referring to applications where some property of the process stream is measured (such as refractive index) and then related to composition of a particular component.

ELECTROANALYTICAL INSTRUMENTS

Conductometric Analysis Solutions of electrolytes in ionizing solvents (e.g., water) conduct current when an electrical potential is applied across electrodes immersed in the solution. Conductance is a function of ion concentration, ionic charge, and ion mobility. Conductance measurements are ideally suited for measurement of the concentration of a single strong electrolyte in dilute solutions. At higher concentrations conductance becomes a complex, nonlinear function of concentration requiring suitable calibration for quantitative measurements.

Measurement of pH The primary detecting element in pH measurement is the glass electrode. A potential is developed at the pH-sensitive glass membrane as a result of differences in hydrogen ion activity in the sample and a standard solution contained within the electrode. This potential measured relative to the potential of the reference electrode gives a voltage which is expressed as pH. Instrumentation for pH measurement is among the most widely used process measurement devices. Rugged electrode systems and highly reliable electronic circuits have been developed for this use.

After installation, the majority of pH measurement problems are sensor-related, mostly on the reference side, including junction plugging, poisoning, and depletion of electrolyte. For the glass (measuring electrode), common difficulties are broken or cracked glass, coating, and etching or abrasion. Symptoms such as drift, sluggish response, unstable readings, and inability to calibrate are indications of measurement problems. Online diagnostics such as impedance measurements, wiring checks, and electrode temperature are now available in most instruments. Other characteristics that can be measured offline include efficiency or slope and asymmetry potential (offset), which indicate whether the unit should be cleaned or changed [Nichols, *Chem. Engr. Prog.*, **90**(12):64, 1994; McMillan, *Chem. Engr. Prog.*, **87**(12):30, 1991].

Specific-Ion Electrodes In addition to the pH glass electrode specific for hydrogen ions, a number of electrodes which are selective for the measurement of other ions have been developed. This selectivity is obtained through the composition of the electrode membrane (glass, polymer, or liquid-liquid) and the composition of the electrode. These electrodes are subject to interference from other ions, and the response is a function of the total ionic strength of the solution. However, electrodes have been designed to be highly selective for specific ions, and when properly used, these provide valuable process measurements.

MOISTURE MEASUREMENT

Moisture measurements are important in the process industries because moisture can foul products, poison reactions, damage equipment, or cause explosions. Moisture measurements include both absolute moisture methods and relative-humidity methods. The absolute methods provide a primary output that can be directly calibrated in terms of dew point temperature, molar concentration, or weight concentration. Loss of weight on heating is the most familiar of these methods. The relative-humidity methods provide a primary output that can be more directly calibrated in terms of percentage of saturation of moisture.

Dew Point Method For many applications the dew point is the desired moisture measurement. When concentration is desired, the relation between water content and dew point is well known and available. The dew point method requires an inert surface whose temperature can be adjusted and measured, a sample gas stream flowing past the surface, a manipulated variable for adjusting the surface temperature to the dew point, and a means of detecting the onset of condensation.

Although the presence of condensate can be detected electrically, the original and most often used method is the optical detection of change in light reflection from an inert metallic-surface mirror. Some instruments measure the attenuation of reflected light at the onset of condensation. Others measure the increase of light dispersed and scattered by the condensate instead of, or in addition to, the reflected-light measurement. Surface cooling is obtained with an expendable refrigerant liquid, conventional mechanical refrigeration, or thermoelectric cooling. Surface-temperature measurement is usually made with a thermocouple or a thermistor.

Piezoelectric Method A piezoelectric crystal in a suitable oscillator circuit will oscillate at a frequency dependent on its mass. If the crystal has a stable hygroscopic film on its surface, the equivalent mass of the crystal varies with the mass of water sorbed in the film. Thus the frequency of oscillation depends on the water in the film. The analyzer contains two such crystals in matched oscillator circuits. Typically, valves alternately direct the sample to one crystal and a dry gas to the other on a 30-s cycle. The oscillator frequencies of the two circuits are compared electronically, and the output is the difference between the two frequencies. This output is then representative of the moisture content of the sample. The output frequency is usually converted to a variable dc voltage for meter readout and recording. Multiple ranges are provided for measurement from about 1 ppm to near saturation. The dry reference gas is preferably the same as the sample except for the moisture content of the sample. Other reference gases which are adsorbed in a manner similar to the dried sample gas may be used. The dry gas is usually supplied by an automatic dryer. The method requires a vapor sample to the detector. Mist striking the detector destroys the accuracy of measurement until it vaporizes or is washed off the crystals. Water droplets or mist may destroy the hygroscopic film, thus requiring crystal replacement. Vaporization or gas-liquid strippers may sometimes be used for the analysis of moisture in liquids.

Capacitance Method Several analyzers utilize the high dielectric constant of water for its detection in solutions. The alternating electric current through a capacitor containing all or part of the sample between the capacitor plates is measured. Selectivity and sensitivity are enhanced by increasing the concentration of moisture in the cell by filling the capacitor sample cell with a moisture-specific sorbent as part of the dielectric. This both increases the moisture content and reduces the amount of other interfering sample components. Granulated alumina is the most frequently used sorbent. These detectors may be cleaned and recharged easily and with satisfactory reproducibility if the sorbent itself is uniform.

Oxide Sensors Aluminum oxide can be used as a sensor for moisture analysis. A conductivity call has one electrode node of aluminum, which is anodized to form a thin film of aluminum oxide, followed by coating with a thin layer of gold (the opposite electrode). Moisture is selectively adsorbed through the gold layer and into the hygroscopic aluminum oxide layer, which in turn determines the electrical conductivity between gold and aluminum oxide. This value can be related to ppm water in the sample. This sensor can operate between near vacuum to several hundred atmospheres, and it is independent of flow rate (including static conditions). Temperature, however, must be carefully monitored. A similar device is based on phosphorous pentoxide. Moisture content influences the electric current between two inert metal electrodes, which are fabricated as a helix on the inner wall of a tubular nonconductive sample cell. For a constant dc voltage applied to the electrodes, a current flows that is proportional to moisture. The moisture is absorbed into the hygroscopic phosphorous pentoxide, where the current electrolyzes the water molecules into hydrogen and oxygen. This sensor will handle moisture up to 1000 ppm and 6-atm pressure. As with the aluminum oxide ion, temperature control is very important.

Photometric Moisture Analysis This analyzer requires a light source, a filter wheel rotated by a synchronous motor, a sample cell, a detector to measure the light transmitted, and associated electronics. Water has two absorption bands in the near-infrared region at 1400 and 1900 nm. This analyzer can measure moisture in liquid or gaseous samples at levels from 5 ppm up to 100 percent, depending on other chemical species in the sample. Response time is less than 1 s, and samples can be run up to 300°C and 400 psig.

OTHER TRANSDUCERS

Other types of transducers used in process measurements include mechanical drivers such as gear trains and electrical drivers such as a differential transformer or a Hall effect (semiconductor-based) sensor.

Gear Train Rotary motion and angular position are easily transduced by various types of gear arrangements. A gear train in conjunction with a mechanical counter is a direct and effective way to obtain a digital readout of shaft rotations. The numbers on the counter can mean anything desired, depending on the gear ratio and the actuating device used to turn the shaft. A pointer attached to a gear train can be used to indicate a number of revolutions or a small fraction of a revolution for any specified pointer rotation.

Differential Transformer These devices produce an ac electrical output from linear movement of an armature. They are very versatile in that they can be designed for a full range of output with any range of armature travel up to several inches. The transformers have one or two primaries and two secondaries connected to oppose each other. With an ac voltage applied to the primary, the output voltage depends on the position of the armature and the coupling. Such devices produce accuracies of 0.5 to 1.0 percent of full scale and are used to transmit forces, pressures, differential pressures, or weights up to 1500 m. They can also be designed to transmit rotary motion.

Hall Effect Sensors Some semiconductor materials exhibit a phenomenon in the presence of a magnetic field which is adaptable to sensing devices. When a current is passed through one pair of wires attached to a semiconductor, such as germanium, another pair of wires properly attached and oriented with respect to the semiconductor will develop a voltage proportional to the magnetic field present and the current in the other pair of wires. Holding the exciting current constant and moving a permanent magnet near the semiconductor produce a voltage output proportional to the movement of the magnet. The magnet may be attached to a process variable measurement device which moves the magnet as the variable changes. Hall effect devices provide high speed of response, excellent temperature stability, and no physical contact.

SAMPLING SYSTEMS FOR PROCESS ANALYZERS

The sampling system consists of all the equipment required to present a process analyzer with a clean representative sample of a process stream and to dispose of that sample. When the analyzer is part of an automatic control loop, the reliability of the sampling system is as important as the reliability of the analyzer or the control equipment. Sampling systems have several functions. The sample must be withdrawn from the process, transported, conditioned, introduced to the analyzer, and disposed. Probably the most common problem in sample system design is the lack of realistic information concerning the properties of the process material at the sampling point. Another common problem is the lack of information regarding the conditioning required so that the analyzer may utilize the sample without malfunction for long periods. Some samples require enough conditioning and treating that the sampling systems become equivalent to miniature online processing plants. These systems possess many of the same fabrication, reliability, and operating problems as small-scale pilot plants except that the sampling system must generally operate reliably for much longer periods of time.

Selecting the Sampling Point The selection of the sampling point is based primarily on supplying the analyzer with a sample whose composition or physical properties are pertinent to the control function to be performed. Other considerations include selecting locations that provide representative homogeneous samples with minimum transport delay, locations which collect a minimum of contaminating material, and locations which are accessible for test and maintenance procedures.

Sample Withdrawal from Process A number of considerations are involved in the design of sample withdrawal devices which will provide representative samples. For example, in a horizontal pipe carrying process fluid, a sample point on the bottom of the pipe will collect a maximum count of rust, scale, or other solid materials being carried along by the process fluid. In a gas stream, such a location will also collect a maximum amount of liquid contaminants. A sample point on the top side of a pipe will, for liquid streams, collect a maximum amount of vapor contaminants being carried along. Bends in the piping which produce swirls or cause centrifugal concentration of the denser phase may cause maximum contamination to be at unexpected locations. Two-phase process materials are difficult to sample for a total-composition representative sample.

A typical method for obtaining a sample of process fluid well away from vessel or pipe walls is an eduction tube inserted through a packing gland. This sampling method withdraws liquid sample and vaporizes it for transporting to the analyzer location. The transport lag time from the end of the probe to the vaporizer is minimized by using tubing having a small internal volume compared with pipe and valve volumes.

This sample probe may be removed for maintenance and reinstalled without shutting down the process. The eduction tube is made of material which will not corrode so that it will slide through the packing gland even after long periods of service. There may be a small amount of process fluid leakage until the tubing is withdrawn sufficiently to close the gate valve. A swaged ferrule on the end of the tube prevents accidental ejection of the eduction tube prior to removal of the packing gland. The section of pipe surrounding the eduction tube and extending into the process vessel provides mechanical protection for the eduction tube.

Sample Transport Transport time—the time elapsed between sample withdrawal from the process and its introduction into the analyzer—should be minimized, particularly if the analyzer is an automatic analyzer-controller. Any sample transport time in the analyzer-controller loop must be treated as equivalent to process dead time in determining conventional feedback controller settings or in evaluating controller performance. Reduction in transport time usually means transporting the sample in the vapor state.

Design considerations for sample lines are as follows:

1. The structural strength or protection must be compatible with the area through which the sample line runs.
2. Line size and length must be small enough to meet transport time requirements without excessive pressure drop or excessive bypass of sample at the analyzer input.
3. Line size and internal surface quality must be adequate to prevent clogging by the contaminants in the sample.
4. The prevention of a change of state of the sample may require installation, refrigeration, or heating of the sample line.
5. Sample line material must be such as to minimize corrosion due to sample or the environment.

Sample Conditioning Sample conditioning usually involves the removal of contaminants or some deleterious component from the sample mixture and/or the adjustment of temperature, pressure, and flow rate of the sample to values acceptable to the analyzer. Some of the more common contaminants which must be removed are rust, scale, corrosion products, deposits due to chemical reactions, and tar. In sampling some process streams, the material to be removed may include the primary process product such as polymer or the main constituent of the stream such as oil. In other cases the material to be removed is present in trace quantities, e.g., water in an online chromatograph sample which can damage the chromatographic column packing. When contaminants or other materials which will hinder analysis represent a large percentage of the stream composition, their removal may significantly alter the integrity of the sample. In some cases, removal must be done as part of the analysis function so that removed material can be accounted for. In other cases, proper calibration of the analyzer output will suffice.

TELEMETERING AND TRANSMISSION

ANALOG SIGNAL TRANSMISSION

Modern control systems permit the measurement device, the control unit, and the final control element to be physically separated by several hundred meters, if necessary. This requires the transmission of the measured variable from the measurement device to the control unit, and the transmission of the controller output from the control unit to the final control element. In each case, transmission of a single value in only one direction is required. Such requirements can be met by analog signal transmission. A measurement range is defined for the value to be transmitted, and the value is basically transmitted as a percent of the span. For the measured variable, the logical span is the measurement span. For the controller output, the logical span is the range of the final control element (e.g., valve fully closed to valve fully open).

For pneumatic transmission systems, the signal range used for the transmission is 3 to 15 psig. In each pneumatic transmission system, there can be only one transmitter, but there can be any number of receivers. When most measurement devices were pneumatic, pneumatic transmission was the logical choice. However, with the displacement of pneumatic measurement devices by electronic devices, pneumatic transmission is now limited to applications where the unique characteristics of pneumatics make it the logical choice.

In order for electronic transmission systems to be less susceptible to interference from magnetic fields, current is used for the transmission signal instead of voltage. The signal range is 4 to 20 mA. In each circuit or "current loop," there can be only one transmitter. There can be more than one receiver, but not an unlimited number. For each receiver, a 250-Ω *range resistor* is inserted into the current loop, which provides a 1- to 5-V input to the receiving device. The number of receivers is limited by the power available from the transmitter.

Both pneumatic and electronic transmission use a *live zero*. This enables the receiver to distinguish a transmitted value of 0 percent of span from a transmitter or transmission system failure. Transmission of 0 percent of span provides a signal of 4 mA in electronic transmission. Should the transmitter or the transmission system fail (i.e., an open circuit in a current loop), the signal level would be 0 mA.

For most measurement variable transmissions, the lower range corresponds to 4 mA and the upper range corresponds to 20 mA. On an open circuit, the measured variable would fail to its lower range. In some applications, this is undesirable. For example, in a fired heater that is heating material to a target temperature, failure of the temperature measurement to its lower range value would drive the output of the combustion control logic to the maximum possible firing rate. In such applications, the analog transmission signal is normally inverted, with the upper range corresponding to 4 mA and the lower range corresponding to 20 mA. On an open circuit, the measured variable would fail to its upper range. For the fired heater, failure of the measured variable to its upper range would drive the output of the combustion control logic to the minimum firing rate.

DIGITAL SYSTEMS

With the advent of the microprocessor, digital technology began to be used for data collection, feedback control, and all other information processing requirements in production facilities. Such systems must acquire data from a variety of measurement devices, and control systems must drive final control elements.

Analog Inputs and Outputs Analog inputs are generally divided into two categories:

1. *High level.* Where the source is a process transmitter, the range resistor in the current loop converts the 4- to 20-mA signal to a 1- to 5-V signal. The conversion equipment can be unipolar (i.e., capable of processing only positive voltages).
2. *Low level.* The most common low-level signals are inputs from thermocouples. These inputs rarely exceed 30 mV and could be zero or even negative. The conversion equipment must be bipolar (i.e., capable of processing positive and negative voltages).

Ultimately, such signals are converted to digital values via an analog-to-digital (A/D) converter. However, the A/D converter is normally preceded by two other components:

1. *Multiplexer.* This permits one A/D converter to service multiple analog inputs. The number of inputs to a multiplexer is usually between 8 and 256.
2. *Amplifier.* As A/D converters require high-level signals, a high-gain amplifier is required to convert low-level signals to high-level signals.

One of the important parameters for the A/D converter is its resolution. The resolution is stated in terms of the number of significant binary digits (bits) in the digital value. As the repeatability of most process transmitters is around 0.1 percent, the minimum acceptable resolution for a bipolar A/D converter is 12 bits, which translates to 11 data bits plus 1 bit for the sign. With this resolution, the analog input values can be represented to 1 part in 2^{11}, or 1 part in 2048. Normally, a 5-V input is converted to a digital value of 2000, which effectively gives a resolution of 1 part in 2000, or 0.05 percent. Very few process control systems utilize resolutions higher than 14 bits, which translates to a resolution of 1 part in 8000, or 0.0125 percent.

For 4- to 20-mA inputs, the resolution is not quite as good as stated above. For a 12-bit bipolar A/D converter, 1 V converts to a digital value of 400. Thus, the range for the digital value is 400 to 2000, making the effective input resolution 1 part in 1600, or 0.0625 percent. Sometimes this is expressed as a resolution of 7.4 bits, where $2^{7.4} = 1600$.

On the output side, dedicated digital-to-analog (D/A) converters are provided for each analog output. Outputs are normally unipolar and require a lower resolution than inputs. A 10-bit resolution is normally sufficient, giving a resolution of 1 part in 1000, or 0.1 percent.

Pulse Inputs Where the sensor within the measurement device is digital, analog-to-digital conversion can be avoided. For rotational devices, the rotational element can be outfitted with a shaft encoder that generates a known number of pulses per revolution. The digital system can process such inputs in any of the following ways:

1. Count the number of pulses over a fixed interval of time.
2. Determine the time for a specified number of pulses.
3. Determine the duration of time between the leading (or trailing) edges of successive pulses.

Of these, the first option is the most commonly used in process applications. Turbine flowmeters are probably the most common example where pulse inputs are used. Another example is a watthour meter. Basically any measurement device that involves a rotational element can be interfaced via pulses. Occasionally, a nonrotational measurement device can generate pulse outputs. One example is the vortex-shedding meter, where a pulse can be generated when each vortex passes over the detector.

Serial Interfaces Some very important measurement devices cannot be reasonably interfaced via either analog or pulse inputs. Two examples are the following:

1. Chromatographs can perform a total composition analysis for a sample. It is possible but inconvenient to provide an analog input for each component. Furthermore, it is often desirable to capture other information, such as the time that the analysis was made (normally the time the sample was injected).
2. Load cells are capable of resolutions of 1 part in 100,000. A/D converters for analog inputs cannot even approach such resolutions.

One approach to interfacing with such devices involves serial interfaces. This has two aspects:

1. *Hardware interface.* The RS-232 interface standard is the basis for most serial interfaces.
2. *Protocol.* This is interpreting the sequence of characters transmitted by the measurement device. There are no standards for protocols, which means that custom software is required.

One advantage of serial interfaces is that two-way communication is possible. For example, a "tare" command can be issued to a load cell.

Microprocessor-Based Transmitters The cost of microprocessor technology has declined to the point where it is economically feasible to incorporate a microprocessor into each transmitter. Such microprocessor-based transmitters are often referred to as *smart* transmitters. As opposed to conventional or *dumb* transmitters, the smart transmitters offer the following capabilities:

1. They check on the internal electronics, such as verifying that the voltage levels of internal power supplies are within specifications.
2. They check on environmental conditions within the instruments, such as verifying that the case temperature is within specifications.
3. They perform compensation of the measured value for conditions within the instrument, such as compensating the output of a pressure transmitter for the temperature within the transmitter. Smart transmitters are much less affected by temperature and pressure variations than are conventional transmitters.
4. They perform compensation of the measured value for other process conditions, such as compensating the output of a capacitance level transmitter for variations in process temperature.
5. They linearize the output of the transmitter. Functions such as square root extraction of the differential pressure for a head-type flowmeter can be done within the instrument instead of within the control system.
6. They configure the transmitter from a remote location, such as changing the span of the transmitter output.
7. They do automatic recalibration of the transmitter. Although this is highly desired by users, the capabilities, if any, in this respect depend on the type of measurement.

Due to these capabilities, smart transmitters offer improved performance over conventional transmitters.

Transmitter/Actuator Networks With the advent of smart transmitters and smart actuators, the limitations of the 4- to 20-mA analog signal transmission retard the full utilization of the capabilities of smart devices. For smart transmitters, the following capabilities are required:

1. *Transmission of more than one value from a transmitter.* Information beyond the measured variable is available from the smart transmitter. For example, a smart pressure transmitter can also report the temperature within its housing. Knowing that this temperature is above normal values permits corrective action to be taken before the device fails. Such information is especially important during the initial commissioning of a plant.
2. *Bidirectional transmission.* Configuration parameters such as span, engineering units, and resolution must be communicated to the smart transmitter.

Similar capabilities are required for smart actuators.

Eventually this technology, generally referred to as field bus, will replace all other forms of transmission (analog, pulse, and serial). Its acceptance by the user community has been slow, mainly due to the absense of a standard. Smart transmitters and smart valves have gained widespread acceptance, but initially the signal transmission continued to be via current loops. During this interim, several manufacturers introduced products based on proprietary communications technologies. Most manufacturers released sufficient information for others to develop products using their communications technologies, but those considering doing so found the costs to be an impediment.

A standard is now available, specifically, IEC 61158, "Digital Data Communications for Measurement and Control." In most industries, a standard defines only one way of doing something, but not in the computer industry. IEC 61158 defines five different communications technologies. Several manufacturers had selected a direction, and standards are not immune to commercial considerations. The net result is that one person's "field bus" might not be the same as another's. Most plants select one technology for their field bus, although it is possible to have multiple field bus interfaces on a control system and not have all the same type.

Although there will be winners and losers, most companies find the appealing features of field bus technologies too great to continue with the older technologies. U.S. manufacturers have generally chosen a field bus technology known as Foundation Fieldbus. In Europe, a field bus technology known as Profibus has gained wide acceptance. Probably the major advantages of these technologies lie in installation and commissioning. There are fewer wires and connections in a field bus installation. In most plants, the cost of the wiring is proving to be a minor issue. With fewer wires and connections, errors in field bus installations are fewer and can be located more quickly. The major cost savings prove to be in configuring and commissioning the system. With time, field bus will largely replace analog, pulse, and serial signal transmissions within industrial facilities.

FILTERING AND SMOOTHING

A signal received from a process transmitter generally contains the following components distinguished by their frequency [frequencies are measured in hertz (Hz), with 60-cycle ac being a 60-Hz frequency):

1. *Low-frequency process disturbances.* The control system is expected to react to these disturbances.
2. *High-frequency process disturbances.* The frequency of these disturbances is beyond the capability of the control system to effectively react.
3. *Measurement noise.*
4. *Stray electrical pickup, primarily 50- or 60-cycle ac.*

The objective of filtering and smoothing is to remove the last three components, leaving only the low-frequency process disturbances. Normally this has to be accomplished by using the proper combination of analog and digital filters. Sampling a continuous signal results in a phenomenon often referred to as aliasing or foldover. To represent a sinusoidal signal, a minimum of four samples are required during each cycle. Consequently, when a signal is sampled at a frequency ω_s, all frequencies higher than $(\pi/2)\omega_s$ cannot be represented at their original frequency. Instead, they are present in the sampled signal with their original amplitude but at a lower-frequency harmonic. Because of the aliasing or foldover issues, a combination of analog and digital filtering is usually required. The sampler (i.e., the A/D converter) must be preceded by an analog filter that rejects those high-frequency components such as stray electrical pickup that would result in foldover when sampled. In commercial products, analog filters are normally incorporated into the input processing hardware by the manufacturer. The software then permits the user to specify digital filtering to remove any undesirable low-frequency components.

On the analog side, the filter is often the conventional resistor-capacitor or *RC* filter. However, other possibilities exist. For example, one type of A/D converter is called an *integrating A/D* because the converter basically integrates the input signal over a fixed interval of time. By making the interval $\frac{1}{60}$ s, this approach provides excellent rejection of any 60-Hz electrical noise.

On the digital side, the input processing software generally provides for smoothing via the exponentially weighted moving average, which is the digital counterpart to the *RC* network analog filter. The smoothing equation is as follows:

$$y_i = \alpha x_i + (1 - \alpha) y_{i-1} \tag{8-107}$$

where x_i = current value of input
y_i = current output from filter

y_{i-1} = previous output from filter
α = filter coefficient

The degree of smoothing is determined by the filter coefficient α, with $\alpha = 1$ being no smoothing and $\alpha = 0$ being infinite smoothing (no effect of new measurements). The filter coefficient α is related to the filter time constant τ_F and the sampling interval Δt by

$$\alpha = 1 - \exp\left(\frac{-\Delta t}{\tau_F}\right) \tag{8-108}$$

or by the approximation

$$\alpha = \frac{\Delta t}{\Delta t + \tau_F} \tag{8-109}$$

Another approach to smoothing is to use the arithmetic moving average, which is represented by

$$y_i = \frac{\sum_{j=1}^{n} x_{i+1-j}}{n} \tag{8-110}$$

The term *moving* is used because the filter software maintains a storage array with the previous n values of the input. When a new value is received, the oldest value in the storage array is replaced with the new value, and the arithmetic average is recomputed. This permits the filtered value to be updated each time a new input value is received.

In process applications, determining τ_F (or α) for the exponential filter and n for the moving average filter is often done merely by observing the behavior of the filtered value. If the filtered value is "bouncing," the degree of smoothing (that is, τ_F or n) is increased. This can easily lead to an excessive degree of filtering, which will limit the performance of any control system that uses the filtered value. The degree of filtering is best determined from the frequency spectrum of the measured input, but such information is rarely available for process measurements.

ALARMS

The purpose of an alarm is to alert the process operator to a process condition that requires immediate attention. An alarm is said to occur whenever the abnormal condition is detected and the alert is issued. An alarm is said to return to normal when the abnormal condition no longer exists. Analog alarms can be defined on measured variables, calculated variables, controller outputs, and the like. For analog alarms, the following possibilities exist:

1. *High/low alarms.* A high alarm is generated when the value is greater than or equal to the value specified for the high-alarm limit. A low alarm is generated when the value is less than or equal to the value specified for the low-alarm limit.
2. *Deviation alarms.* An alarm limit and a target are specified. A high-deviation alarm is generated when the value is greater than or equal to the target plus the deviation alarm limit. A low-deviation alarm is generated when the value is less than or equal to the target minus the deviation alarm limit.
3. *Trend or rate-of-change alarms.* A limit is specified for the maximum rate of change, usually specified as a change in the measured value per minute. A high-trend alarm is generated when the rate of change of the variable is greater than or equal to the value specified for the trend alarm limit. A low-trend alarm is generated when the rate of change of the variable is less than or equal to the negative of the value specified for the trend alarm limit.

Most systems permit multiple alarms of a given type to be configured for a given value. For example, configuring three high alarms provides a high alarm, a high-high alarm, and a high-high-high alarm.

One operational problem with analog alarms is that noise in the variable can cause multiple alarms whenever its value approaches a limit. This can be avoided by defining a dead band on the alarm. For example, a high alarm would be processed as follows:

1. *Occurrence.* The high alarm is generated when the value is greater than or equal to the value specified for the high-alarm limit.
2. *Return to normal.* The high-alarm return to normal is generated when the value is less than or equal to the high-alarm limit less the dead band.

As the degree of noise varies from one input to the next, the dead band must be individually configurable for each alarm.

Discrete alarms can be defined on discrete inputs, limit switch inputs from on/off actuators, and so on. For discrete alarms, the following possibilities exist:

1. *Status alarms.* An expected or normal state is specified for the discrete value. A status alarm is generated when the discrete value is other than its expected or normal state.
2. *Change-of-state alarm.* A change-of-state alarm is generated on any change of the discrete value.

The expected sequence of events on an alarm is basically as follows:

1. The alarm occurs. This usually activates an audible annunciator.
2. The alarm occurrence is acknowledged by the process operator. When all alarms have been acknowledged, the audible annunciator is silenced.
3. Corrective action is initiated by the process operator.
4. The alarm condition returns to normal.

However, additional requirements are imposed at some plants. Sometimes the process operator must acknowledge the alarm's return to normal. Some plants require that the alarm occurrence be reissued if the alarm remains in the occurred state longer than a specified time. Consequently, some "personalization" of the alarming facilities is done.

When alarms were largely hardware-based (i.e., the panel alarm systems), the purchase and installation of the alarm hardware imposed a certain discipline on the configuration of alarms. With digital systems, the suppliers have made it extremely easy to configure alarms. In fact, it is sometimes easier to configure alarms on a measured value than not to configure the alarms. Furthermore, the engineer assigned the responsibility for defining alarms is often paranoid that an abnormal process condition will go undetected because an alarm has not been configured. When alarms are defined on every measured and calculated variable, the result is an excessive number of alarms, most of which are duplicative and unnecessary.

The accident at the Three Mile Island nuclear plant clearly demonstrated that an alarm system can be counterproductive. An excessive number of alarms can distract the operator's attention from the real problem that needs to be addressed. Alarms that merely tell the operator something that is already known do the same. In fact, a very good definition of a nuisance alarm is one that informs the operator of a situation of which the operator is already aware. The only problem with applying this definition lies determining what the operator already knows.

Unless some discipline is imposed, engineering personnel, especially where contractors are involved, will define far more alarms than plant operations require. This situation may be addressed by simply setting the alarm limits to values such that the alarms never occur. However, changes in alarms and alarm limits are changes from the perspective of the process safety management regulations. It is prudent to impose the necessary discipline to avoid an excessive number of alarms. Potential guidelines are as follows:

1. For each alarm, a specific action is expected from the process operator. Operator actions such as call maintenance are inappropriate with modern systems. If maintenance needs to know, modern systems can inform maintenance directly.
2. Alarms should be restricted to abnormal situations for which the process operator is responsible. A high alarm on the temperature in one of the control system cabinets should not be issued to the process operator. Correcting this situation is the responsibility of maintenance, not the process operator.
3. Process operators are expected to be exercising normal surveillance of the process. Therefore, alarms are not appropriate for situations known to the operator either through previous alarms or through normal process surveillance. The "sleeping operator" problem can be addressed by far more effective means than the alarm system.
4. When the process is operating normally, no alarms should be triggered. Within the electric utility industry, this design objective is

known as *darkboard*. Application of darkboard is especially important in batch plants, where much of the process equipment is operated intermittently.

Ultimately, guidelines such as those above will be taken seriously only if production management pays attention to the process alarms. The consequences of excessive and redundant alarms will be felt primarily by those responsible for production operations. Therefore, production management must make adequate resources available for reviewing and analyzing the proposed alarm configurations.

Another serious distraction to a process operator is the multiple-alarm event, where a single event within the process results in multiple alarms. When the operator must individually acknowledge each alarm, considerable time can be lost in silencing the obnoxious annunciator before the real problem is addressed. Air-handling systems are especially vulnerable to this, where any fluctuation in pressure (e.g., resulting from a blower trip) can cause a number of pressure alarms to occur.

Point alarms (high alarms, low alarms, status alarms, etc.) are especially vulnerable to the multiple-alarm event. This can be addressed in one of two ways:

1. *Ganging alarms.* Instead of individually issuing the point alarms, all alarms associated with a certain aspect of the process are configured to give a single trouble alarm. The responsibility rests entirely with the operator to determine the nature of the problem.

2. *Intelligent alarms.* Logic is incorporated into the alarm system to determine the nature of the problem and then issue a single alarm to the process operator.

While the intelligent alarm approach is clearly preferable, substantial process analysis is required to support intelligent alarming. Meeting the following two objectives is quite challenging:

1. The alarm logic must consistently detect abnormal conditions within the process.

2. The alarm logic must not issue an alert to an abnormal condition when in fact none exists.

Often the latter case is more challenging than the former. Logically, the intelligent alarm effort must be linked to the process hazards analysis. Developing an effective intelligent alarming system requires substantial commitments of effort, involving process engineers, control systems engineers, and production personnel. Methodologies such as expert systems can facilitate the implementation of an intelligent alarming system, but they must still be based on a sound analysis of the potential process hazards.

DIGITAL TECHNOLOGY FOR PROCESS CONTROL

General References: Auslander and Ridgely, *Design and Implementation of Real-Time Software for the Control of Mechanical Systems*, Prentice-Hall, Upper Saddle River, N.J., 2002. Herb, *Understanding Distributed Processor Systems for Control*, ISA, Research Triangle Park, N.C., 1999. Hughes, *Programmable Controllers*, ISA, Research Triangle Park, N.C., 1997. Johnson, *Process Control Instrumentation Technology*, 6th ed., Prentice-Hall, Upper Saddle River, N.J., 2000. Liptak, *Instrument Engineers Handbook*, Chilton Book Company, Philadelphia, 1995. Webb and Reis, *Programmable Logic Controllers*, 4th ed., Prentice-Hall, Upper Saddle River, N.J., 2002.

Since the 1970s, process controls have evolved from pneumatic analog technology to electronic analog technology to microprocessor-based controls. Electronic and pneumatic controllers have now virtually disappeared from process control systems, which are dominated by programmable electronic systems based on microprocessor technology.

HIERARCHY OF INFORMATION SYSTEMS

Coupling digital controls with networking technology permits information to be passed from level to level within a corporation at high rates of speed. This technology is capable of presenting the measured variable from a flow transmitter installed in a plant in a remote location anywhere in the world to the company headquarters in less than 1 s.

A hierarchical representation of the information flow within a company leads to a better understanding of how information is passed from one layer to the next. Such representations can be developed in varying degrees of detail, and most companies have developed one that describes their specific practices. The following hierarchy consists of five levels, as shown in Fig. 8-30.

Measurement Devices and Final Control Elements This lowest layer couples the control and information systems to the process. The measurement devices provide information on the current conditions within the process. The final control elements permit control decisions to be imposed on the process. Although traditionally analog, smart transmitters and smart valves based on microprocessor technology are now beginning to dominate this layer.

Safety and Environmental/Equipment Protection The level 2 functions play a critical role by ensuring that the process is operating safely and satisfies environmental regulations. Process safety relies on the principle of multiple protection layers that involve groupings of equipment and human actions. One layer includes process control functions, such as alarm management during abnormal situations, and safety instrumented systems for emergency shutdowns. The safety equipment (including sensors and block valves) operates independently of the regular instrumentation used for regulatory control in level 3. Sensor validation techniques can be employed to confirm that the sensors are functioning properly.

Regulatory Controls The objective of this layer is to operate the process at or near the targets supplied by a higher layer in the hierarchy. To achieve consistent process operations, a high degree of automatic control is required from the regulatory layer. The direct result is a reduction in variance in the key process variables. More uniform product quality is an obvious benefit. However, consistent process operation is a prerequisite for optimizing the process operations. To ensure success for the upper-level functions, the first objective of any automation effort must be to achieve a high degree of regulatory control.

Real-Time Optimization Determining the most appropriate targets for the regulatory layer is the responsibility of the RTO layer. Given the current production and quality targets for a unit, RTO determines how the process can be best operated to meet them. Usually this optimization has a limited scope, being confined to a single production unit or possibly even a single unit operation within a production unit. RTO translates changes in factors such as current process efficiencies, current energy costs, cooling medium temperatures, and so on to changes in process operating targets so as to optimize process operations.

Production Controls The nature of the production control logic differs greatly between continuous and batch plants. A good example of production control in a continuous process is refinery optimization. From the assay of the incoming crude oil, the values of the various possible refined products, the contractual commitments to deliver certain products, the performance measures of the various units within a refinery, and the like, it is possible to determine the mix of products that optimizes the economic return from processing this crude. The solution of this problem involves many relationships and constraints and is solved with techniques such as linear programming.

In a batch plant, production control often takes the form of routing or short-term scheduling. For a multiproduct batch plant, determining the long-term schedule is basically a manufacturing resource planning (MRP) problem, where the specific products to be manufactured and the amounts to be manufactured are determined from the outstanding orders, the raw materials available for production, the production capacities of the process equipment, and other factors. The goal of the MRP effort is the long-term schedule, which is a list of the products to be manufactured over a specified period of time (often one week). For each product on the list, a target amount is also specified. To manufacture this amount usually involves several batches. The term *production run* often refers to the sequence of batches required to make the target amount of product, so in effect the long-term schedule is a list of production runs.

Most multiproduct batch plants have more than one piece of equipment of each type. Routing refers to determining the specific pieces of equipment that will be used to manufacture each run on the long-term production schedule. For example, the plant might have five reactors, eight neutralization tanks, three grinders, and four packing machines. For a given run, a rather large number of routes are possible. Furthermore, rarely is only one run in progress at a given time. The objective of routing is to determine the specific pieces of production equipment to be used for each run on the long-term production schedule. Given the dynamic nature of the production process (equipment failures, insertion/deletion of runs into the long-term schedule, etc.), the solution of the routing problem continues to be quite challenging.

Corporate Information Systems Terms such as *management information systems* (MIS), *enterprise resource planning* (ERP), *supply chain management* (SCM), and *information technology* (IT) are frequently used to designate the upper levels of computer systems within a corporation. From a control perspective, the functions performed at this level are normally long-term and/or strategic. For example, in a processing plant, long-term contracts are required with the providers of the feedstocks. A forecast must be developed for the demand for possible products from the plant. This demand must be translated to needed raw materials, and then contracts executed with the suppliers to deliver these materials on a relatively uniform schedule.

DIGITAL HARDWARE IN PROCESS CONTROL

Digital control technology was first applied to process control in 1959, using a single central computer (and analog backup for reliability). In the mid-1970s, a microcomputer-based process control architecture referred to as a *distributed control system (DCS)* was introduced and rapidly became a commercial success. A DCS consists of some number of microprocessor-based nodes that are interconnected by a digital communications network, often called a data highway. Today the DCS is still dominant, but there are other options for carrying out computer control, such as single-loop controllers, programmable logic controllers, and personal computer controllers. A brief review of each type of controller device is given below; see the following section "Controllers, Final Control Elements, and Regulators" for more details on controller hardware options.

Single-Loop Controllers The single-loop controller (SLC) is the digital equivalent of analog single-loop controllers. It is a self-contained microprocessor-based unit that can be rack-mounted. Many manufacturers produce single processor units that handle cascade control or multiple loops, typically 4, 8, or 16 loops per unit, and incorporate self-tuning or auto-tuning PID control algorithms.

Programmable Logic Controllers Programmable logic controllers (PLCs) are simple digital devices that are widely used to control sequential and batch processes, although they now have additional functions that implement PID control and other mathematical operations. PLCs can be utilized as stand-alone devices or in conjunction with digital computer control systems. Because the logical functions are stored in main memory, one measure of a PLC's capability is its memory scan rate. Most PLCs are equipped with an internal timing capability to delay an action by a prescribed amount of time, to execute an action at a prescribed time, and so on. Newer PLC models often are networked to serve as one component of a DCS, with operator I/O provided by a separate component in the network. A distinction is made between configurable and programmable PLCs. The term *configurable* implies that logical operations (performed on inputs to yield a desired output) are located in PLC memory, perhaps in the form of ladder diagrams by selecting from a PLC menu or by direct interrogation of the PLC. Most control engineers prefer the simplicity of configuring the PLC to the alternative of programming it. However, some batch applications, particularly those involving complex sequencing, are best handled by a programmable approach.

Personal Computer Controllers In comparison with PLCs, PCs have the advantages of lower purchase cost, graphics output, large memory, large selection of software products (including databases and development tools), more programming options (use of C or Java versus ladder logic), richer operating systems, and open networking. PLCs have the following advantages: lower maintenance cost, operating system and hardware optimized for control, fast boot times, ruggedness, low failure rate, longer support for product models, and self-contained units.

A number of vendors have introduced so-called scalable process control systems. Scalable means that the size of the control and instrumentation systems is easily expanded by simply adding more devices. This feature is possible because of the trend toward more openness (i.e., "plug and play" between devices), smaller size, lower cost, greater flexibility, and more off-the-shelf hardware and software in digital control systems. A typical system includes personal computers, an operating system, object-oriented database technology, modular field-mounted controllers, and plug-and-play integration of both system and intelligent field devices. New devices are automatically recognized and configured with the system. Advanced control algorithms can be executed at the PC level.

Distributed Control System Figure 8-65 depicts a representative distributed control system. The DCS consists of many commonly

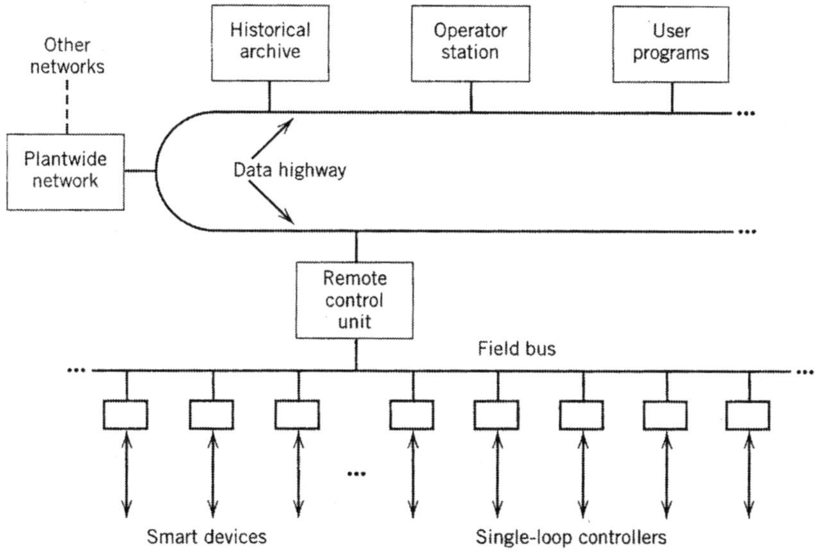

FIG. 8-65 A DCS using a broadband (high-bandwidth) data highway and field bus connected to a single remote control unit that operates smart devices and single-loop controllers.

used components, including multiplexers (MUXs), single-loop and multiple-loop controllers, PLCs, and smart devices. A system includes some of or all the following components:

1. *Control network.* The control network is the communication link between the individual components of a network. Coaxial cable and, more recently, fiber-optic cable have often been used, sometimes with Ethernet protocols. A redundant pair of cables (dual redundant highway) is normally supplied to reduce the possibility of link failure.

2. *Workstations.* Workstations are the most powerful computers in the system, capable of performing functions not normally available in other units. A workstation acts both as an arbitrator unit to route internodal communications and as the database server. An operator interface is supported, and various peripheral devices are coordinated through the workstations. Computationally intensive tasks, such as real-time optimization or model predictive control, are implemented in a workstation. Operators supervise and control processes from these workstations. Operator stations may be connected directly to printers for alarm logging, printing reports, or process graphics.

3. *Remote control units (RCUs).* These components are used to implement basic control functions such as PID control. Some RCUs may be configured to acquire or supply set points to single-loop controllers. Radio telemetry (wireless) may be installed to communicate with MUX units located at great distances.

4. *Application stations.* These separate computers run application software such as databases, spreadsheets, financial software, and simulation software via an OPC interface. OPC is an acronym for object linking and embedding for process control, a software architecture based on standard interfaces. These stations can be used for e-mail and as web servers, for remote diagnosis configuration, and even for operation of devices that have an IP (Internet Protocol) address. Applications stations can communicate with the main database contained in online mass storage systems. Typically hard disk drives are used to store active data, including online and historical databases and nonmemory resident programs. Memory resident programs are also stored to allow loading at system start-up.

5. *Field buses and smart devices.* An increasing number of field-mounted devices are available that support digital communication of the process I/O in addition to, or in place of, the traditional 4- to 20-mA current signal. These devices have greater functionality, resulting in reduced setup time, improved control, combined functionality of separate devices, and control valve diagnostic capabilities. Digital communication also allows the control system to become completely distributed where, e.g., a PID control algorithm could reside in a valve positioner or in a sensor/transmitter.

DISTRIBUTED DATABASE AND THE DATABASE MANAGER

A database is a centralized location for data storage. The use of databases enhances system performance by maintaining complex relations between data elements while reducing data redundancy. A database may be built based on the relational model, the entity relationship model, or some other model. The database manager is a system utility program or programs acting as the gatekeeper to the databases. All functions retrieving or modifying data must submit a request to the manager. Information required to access the database includes the tag name of the database entity, often referred to as a point, the attributes to be accessed, and the values, if modifying. The database manager maintains the integrity of the databases by executing a request only when not processing other conflicting requests. Although a number of functions may read the same data item at the same time, writing by a number of functions or simultaneous read and write of the same data item is not permitted.

To allow flexibility, the database manager must also perform point addition or deletion. However, the ability to create a point type or to add or delete attributes of a point type is not normally required because, unlike other data processing systems, a process control system normally involves a fixed number of point types and related attributes. For example, analog and binary input and output types are required for process I/O points. Related attributes for these point types include tag names, values, and hardware addresses. Different system manufacturers may define different point types using different data structures. We will discuss other commonly used point types and attributes as they appear.

Data Historian A historical database is built similar to an online database. Unlike their online counterparts, the information stored in a historical database is not normally accessed directly by other subsystems for process control and monitoring. Periodic reports and long-term trends are generated based on the archived data. The reports are often used for planning and system performance evaluations such as statistical process (quality) control. The trends may be used to detect process drifts or to compare process variations at different times.

The historical data are sampled at user-specified intervals. A typical process plant contains a large number of data points, but it is not feasible to store data for all points at all times. The user determines if a data point should be included in the list of archive points. Most systems provide archive-point menu displays. The operators are able to add or delete data points to the archive point lists. The sampling periods are normally some multiples of their base scan frequencies. However, some systems allow historical data sampling of arbitrary intervals. This is necessary when intermediate virtual data points that do not have the scan frequency attribute are involved. The archive point lists are continuously scanned by the historical database software. Online databases are polled for data, and the times of data retrieval are recorded with the data obtained. To conserve storage space, different data compression techniques are employed by various manufacturers.

DIGITAL FIELD COMMUNICATIONS AND FIELD BUS

Microprocessor-based equipment, such as smart instruments and single-loop controllers with digital communications capability, are now used extensively in process plants. A field bus, which is a low-cost protocol, is necessary to perform efficient communication between the DCS and devices that may be obtained from different vendors. Figure 8-65 illustrates a LAN-based DCS with field buses and smart devices connected to a data highway.

Presently, there are several regional and industry-based field bus standards, including the French standard (FIP), the German standard (Profibus), and proprietary standards by DCS vendors, generally in the United States, led by the Fieldbus Foundation, a not-for-profit corporation. As of 2002, international standards organizations had adopted all these field bus standards rather than a single unifying standard. However, there will likely be further developments in field bus standards in the future. A benefit of standardizing the field bus is that it has encouraged third-party traditional equipment manufacturers to enter the smart equipment market, resulting in increased competition and improved equipment quality.

Several manufacturers have made available field bus controllers that reside in the final control element or measurement transmitter. A suitable communications modem is present in the device to interface with a proprietary PC-based, or hybrid analog/digital bus network. At present, field bus controllers are single-loop controllers containing 8- and 16-bit microprocessors that support the basic PID control algorithm as well as other functionalities. Case studies in implementing such digital systems have shown significant reductions in cost of installation (mostly cabling and connections) versus traditional analog field communication.

A general movement has also begun in the direction of using the high-speed Ethernet standard (100 Mbit/s or higher), allowing data transfer by TCP/IP that is used pervasively in computer networking. This would allow any smart device to communicate directly with others in the network or to be queried by the operator regarding its status and settings.

INTERNODAL COMMUNICATIONS

For a group of computers to become a network, intercomputer communication is required. Prior to the 1980s, each system vendor used a proprietary protocol to network its computers. Ad hoc approaches were sometimes used to connect third-party equipment but were not cost-effective with regard to system maintenance, upgrade, and expansion.

The recent introduction of standardized communication protocols has led to a decrease in capital cost. Most current DCS network protocol designs are based on the ISO-OSI seven-layer model with physical, data link, network, transport, session, presentation, and application layers.

An effort in standardizing communication protocols for plant automation was initiated by General Motors in the early 1980s. This work culminated in the Manufacturing Automation Protocol (MAP), which adopted the ISO-OSI standards as its basis. MAP specifies a broadband backbone local-area network (LAN) that incorporates a selection of existing standard protocols suitable for discrete component manufacturing. MAP was intended to address the integration of DCSs used in process control. Subsequently, TCP/IP (Transmission Control Protocol/Internet Protocol) was adopted for communication between nodes that have different operating systems.

Communication programs also act as links to the database manager. When data are requested from a remote node, the database manager transfers the request to the remote node database manager via the communication programs. The remote node communication programs then relay the request to the resident database manager and return the requested data. The remote database access and the existence of communications equipment and software are transparent to the user.

PROCESS CONTROL LANGUAGES

Originally, software for process control utilized high-level programming languages such as FORTRAN and BASIC. Some companies have incorporated libraries of software routines for these languages, but others have developed specialty languages characterized by natural language statements. The most widely adopted user-friendly approach is the fill-in-the-forms or table-driven process control languages (PCLs). Typical PCLs include function block diagrams, ladder logic, and programmable logic. The core of these languages is a number of basic function blocks or software modules, such as analog in, digital in, analog out, digital out, PID, summer, and splitter. Using a module is analogous to calling a subroutine in conventional FORTRAN or C programs.

In general, each module contains one or more inputs and an output. The programming involves connecting outputs of function blocks to inputs of other blocks via the graphical-user interface. Some modules may require additional parameters to direct module execution. Users are required to fill in templates to indicate the sources of input values, the destinations of output values, and the parameters for forms/tables prepared for the modules. The source and destination blanks may specify process I/O channels and tag names when appropriate. To connect modules, some systems require filling in the tag names of modules originating or receiving data.

Many DCSs allow users to write custom code (similar to BASIC) and attach it to data points, so that the code is executed each time the point is scanned. The use of custom code allows many tasks to be performed that cannot be carried out by standard blocks.

All process control languages contain PID control blocks of different forms. Other categories of function blocks include

1. *Logical operators.* AND, OR, and exclusive OR (XOR) functions.
2. *Calculations.* Algebraic operations such as addition, multiplication, square root extraction, or special function evaluation.
3. *Selectors.* Min and max functions, transferring data in a selected input to the output or the input to a selected output.
4. *Comparators.* Comparison of two analog values and transmission of a binary signal to indicate whether one analog value exceeds the other.
5. *Timers.* Delayed activation of the output for a programmed duration after activation by the input signal.
6. *Process dynamics.* Emulation of a first-order process lag (or lead) and time delay.

CONTROLLERS, FINAL CONTROL ELEMENTS, AND REGULATORS

General References: Driskell, *Control-Valve Selection and Sizing*, ISA, 1983. McMillan, *Process/Industrial Instruments and Controls Handbook*, 5th ed., McGraw-Hill, 1999. Hammitt, *Cavitation and Multiphase Flow Phenomena*, McGraw-Hill, 1980. ANSI/ISA-75.25.01, *Test Procedure for Control Valve Response Measurement from Step Inputs*, The Instrumentation Systems and Automation Society, 2000. Norton and Karczub, *Fundamentals of Noise and Vibration Analysis for Engineers*, 2d ed., Cambridge University Press, 2003. Ulanski, *Valve and Actuator Technology*, McGraw-Hill, 1991. Kinsler et al., *Fundamentals of Acoustics*, 4th ed., Wiley. *National Electrical Code Handbook*, 9th ed., National Fire Protection Association, Inc.

External control of the process is achieved by devices that are specially designed, selected, and configured for the intended process control application. The text that follows covers three very common function classifications of process control devices: controllers, final control elements, and regulators.

The process controller is the "master" of the process control system. It accepts a set point and other inputs and generates an output or outputs that it computes from a rule or set of rules that is part of its internal configuration. The controller output serves as an input to another controller or, more often, as an input to a final control element. The final control element typically is a device that affects the flow in the piping system of the process. The final control element serves as an interface between the process controller and the process. Control valves and adjustable speed pumps are the principal types discussed.

Regulators, though not controllers or final control elements, perform the combined function of these two devices (controller and final control element) along with the measurement function commonly associated with the process variable transmitter. The uniqueness, control performance, and widespread usage of the regulator make it deserving of a functional grouping of its own.

PNEUMATIC, ELECTRONIC, AND DIGITAL CONTROLLERS

Pneumatic Controllers The pneumatic controller is an automatic controller that uses variable air pressure as input and output signals. An air supply is also required to "power" the mechanical components of the controller and provide an air source for the controller output signal. Pneumatic controllers were first available in the early 1940s but are now rarely used for large-scale industrial control projects. Pneumatic controllers are still used where cost, ruggedness, or the installation requires an all-pneumatic solution.

Pneumatic process transmitters are used to produce a pressure signal that is proportional to the calibrated range of the measuring element. Of the transmitter range 0 to 100 percent is typically represented by a 0.2- to 1.0-bar (3- to 15-psig) pneumatic signal. This signal is sent through tubing to the pneumatic controller process variable feedback connection. The process variable feedback can also be sensed directly in cases where the sensing element has been incorporated into the controller design. Controllers with integral sensing elements are available that sense pressure, differential pressure, temperature, and level.

The pneumatic controller is designed so that 0 to 100 percent output is also represented by 0.2 to 1.0 bar (3 to 15 psig). The output signal is sent through tubing to the control valve or other final control element. Most pneumatic controllers provide a manual control mode where the output pressure is manually set by operating personnel. The controller design also provides a mechanism to adjust the set point.

Early controller designs required "balancing" of the controller output prior to switching to or from automatic and manual modes. This procedure minimized inadvertent disturbance to the process caused by potentially large differences between the automatic and manual output levels. Later designs featured "bumpless" or "procedureless" automatic-to-manual transfer.

Although the pneumatic controller is often used in single-loop control applications, cascade strategies can be implemented where the controller design supports input of external or remote set-point signals. A balancing procedure is typically required to align the remote set point with the local set point before the controller is switched into cascade mode.

Almost all pneumatic controllers include indicators for process variable, set point, and output. Many controller designs also feature integral chart recorders. There are versions of the pneumatic controller that support combinations of proportional, integral, and derivative actions.

The pneumatic controller can be installed into panel boards that are adjacent to the process being controlled or in a centrally located control room. Field-mountable controllers can be installed directly onto the control valve, a nearby pipestand, or wall in close proximity to the control valve and/or measurement transmitter.

If operated on clean, dry plant air, pneumatic controllers offer good performance and are extremely reliable. In many cases, however, plant air is neither clean nor dry. A poor-quality air supply will cause unreliable performance of pneumatic controllers, pneumatic field measurement devices, and final control elements. The main shortcoming of the pneumatic controller is its lack of flexibility when compared to modern electronic controller designs. Increased range of adjustability, choice of alternative control algorithms, the communication link to the control system, and other features and services provided by the electronic controller make it a superior choice in most of today's applications. Controller performance is also affected by the time delay induced by pneumatic tubing runs. For example, a 100-m run of 6.35-mm ($\frac{1}{4}$-in) tubing will typically cause 5 s of apparent process dead time, which will limit the control performance of fast processes such as flows and pressures.

Pneumatic controllers continue to be used in areas where it would be hazardous to use electronic equipment, such as locations with flammable or explosive atmospheres or other locations where compressed air is available but where access to electrical services is limited or restricted.

Electronic (Digital) Controllers Almost all the electronic process controllers used today are microprocessor-based, digital devices. In the transition from pneumatic to electronic controllers, a number of analog controller designs were available. Due to the inflexible nature of the analog designs, these controllers have been almost completely replaced by digital designs. The microprocessor-based controllers contain, or have access to, input/output (I/O) interface electronics that allow various types of signals to enter and leave the controller's processor.

The resolution of the analog I/O channels of the controller varies by manufacturer and age of the design. The 12- to 14-bit conversion resolution of the analog input channels is quite common. Conversion resolution of the analog output channels is typically 10- to 12-bit. Newer designs support up to 16 bit input and output resolution. Although 10-bit output resolution had been considered satisfactory for many years, it has recently been identified as a limitation of control performance. This limitation has emerged as the performance of control valve actuators has improved and the use of other high-resolution field devices, such as variable-speed pump drives, has become more prevalent. These improvements have been driven by the need to deliver higher operating efficiencies and improved product specifications through enhanced process control performance.

Sample rates for the majority of digital controllers are adjustable and range from 1 sample every 5 s to 10 samples per second. Some controller designs have fixed sample rates that fall within the same range. Hardwired low-pass filters are usually installed on the analog inputs to the controller to help protect the sampler from aliasing errors.

The real advantage of digital controllers is the substantial flexibility offered by a number of different configuration schemes. The simplest form of configuration is seen in a controller design that features a number of user-selectable control strategies. These strategies are customized by setting "tunable" parameters within the strategy. Another common configuration scheme uses a library of function blocks that can be selected and combined to form the desired control strategy. Each function block has adjustable parameters. Additional configuration schemes include text-based scripting languages, higher-level languages such as Basic or C, and ladder logic.

Some digital controller designs allow the execution rates of control strategy elements to be set independently of each other and independently of the I/O subsystem sample rate. Data passed from control element to subsystems that operate at slower sample or execution rates present additional opportunities for timing and aliasing errors.

Distributed Control Systems Some knowledge of the distributed control system (DCS) is useful in understanding electronic controllers. A DCS is a process control system with sufficient performance to support large-scale, real-time process applications. The DCS has (1) an operations workstation with input devices, such as a keyboard, mouse, track ball, or other similar device, and a display device, such as a CRT or LCD panel; (2) a controller subsystem that supports various types of controllers and controller functions; (3) an I/O subsystem for converting analog process input signals to digital data and digital data back to analog output signals; (4) a higher-level computing platform for performing process supervision, historical data trending and archiving functions, information processing, and analysis; and (5) communication networks to tie the DCS subsystems, plant areas, and other plant systems together. The component controllers used in the controller subsystem portion of the DCS can be of various types and include multiloop controllers, programmable logic controllers, personal computer controllers, single-loop controllers, and field bus controllers. The type of electronic controller utilized depends on the size and functional characteristic of the process application being controlled. Personal computers are increasingly being used as DCS operations workstations or interface stations in place of custom-built machines. This is due to the low cost and high performance of the PC. See the earlier section "Digital Technology for Process Control."

Multiloop Controllers The multiloop controller is a DCS network device that uses a single 32-bit microprocessor to provide control functions to many process loops. The controller operates independently of the other devices on the DCS network and can support from 20 to 500 loops. Data acquisition capability for 1000 analog and discrete I/O channels or more can also be provided by this controller. The I/O is typically processed through a subsystem that is connected to the controller through a dedicated bus or network interface. The multiloop controller contains a variety of function blocks (for example, PID, totalizer, lead/lag compensator, ratio control, alarm, sequencer, and boolean) that can be "soft-wired" together to form complex control strategies. The multiloop controller also supports additional configuration schemes including text-based scripting languages, higher-level languages such as Basic or C, and, to a limited extent, ladder logic. The multiloop controller, as part of a DCS, communicates with other controllers and human/machine interface (HMI) devices also on the DCS network.

Programmable Logic Controllers The programmable logic controller (PLC) originated as a solid-state, and far more flexible, replacement for the hardwired relay control panel and was first used in the automotive industry for discrete manufacturing control. Today, PLCs are primarily used to implement boolean logic functions, timers, and counters. Some PLCs offer a limited number of math functions and PID control. PLCs are often used with on/off input and output devices such as limit or proximity switches, solenoid-actuated process control valves, and motor switch gear. PLCs vary greatly in size with the smallest supporting less than 128 I/O channels and the largest supporting more than 1023 I/O channels. Very small PLCs combine processor, I/O, and communications functions into a single, self-contained unit. For larger PLC systems, hardware modules such as the power supply, processor module, I/O modules, communication module, and backplane are specified based on the application. These systems support multiple I/O backplanes that can be chained together to increase the I/O count available to the processor. Discrete I/O modules are available that support high-current motor loads and general-purpose voltage and current loads. Other modules support analog I/O and special-purpose I/O for servomotors, stepping motors, high-speed pulse counting, resolvers, decoders, displays, and keyboards. PLC I/O modules often come with indicators to determine the status of key I/O channels. When used as an alternative to a DCS, the PLC is programmed with a handheld or computer-based loader. The PLC is typically programmed with ladder logic or a high-level computer language such as BASIC, FORTRAN, or C. Programmable logic controllers use 16- or 32-bit microprocessors and offer some form of point-to-point serial communications such as RS-232C, RS-485, or networked communication such as Ethernet with proprietary or open protocols. PLCs typically execute the boolean or ladder logic configuration at high rates; 10-ms execution intervals are common. This does not necessarily imply that the analog I/O or PID control functions are executed at the same rate. Many PLCs execute the analog program at a much slower rate. Manufacturers' specifications must be consulted.

Personal Computer Controller Because of its high performance at low cost and its unexcelled ease of use, application of the personal computer (PC) as a platform for process controllers is growing. When configured to perform scan, control, alarm, and data acquisition (SCADA) functions and combined with a spreadsheet or database management application, the PC controller can be a low-cost, basic alternative to the DCS or PLC. Using the PC for control requires installation of a board into the expansion slot in the computer, or the PC can be connected to an external I/O module by using a standard communication port on the PC. The communication is typically achieved through a serial interface (RS-232, RS-422, or IEEE-488), universal serial bus (USB), or Ethernet. The controller card/module supports 16- or 32-bit microprocessors. Standardization and high volume in the PC market have produced a large selection of hardware and software tools for PC controllers.

The PC can also be interfaced to a DCS to perform advanced control or optimization functions that are not available within the standard DCS function library.

Single-Loop Controller The single-loop controller (SLC) is a process controller that produces a single output. SLCs can be pneumatic, analog electronic, or microprocessor-based. Pneumatic SLCs are discussed in the pneumatic controller section, and analog electronic SLC is not discussed because it has been virtually replaced by the microprocessor-based design. The microprocessor-based SLC uses an 8- or 16-bit microprocessor with a small number of digital and analog process input channels with control logic for the I/O incorporated within the controller. Analog inputs and outputs are available in the standard ranges (1 to 5 V dc and 4 to 20 mA dc). Direct process inputs for temperature sensors (thermistor RTD and thermocouple types) are available. Binary outputs are also available. The face of the SLC has some form of visible display and pushbuttons that are used to view or adjust control values and configuration. SLCs are available for mounting in panel openings as small as 48 × 48 mm (1.9 × 1.9 in).

The processor-based SLC allows the user to select from a set of predefined control strategies or to create a custom control strategy by using a set of control function blocks. Control function blocks include PID, on/off, lead/lag, adder/subtractor, multiply/divide, filter functions, signal selector, peak detector, and analog track. SLCs feature auto/manual transfer switching, multi-set-point self-diagnostics, gain scheduling, and perhaps also time sequencing. Most processor-based SLCs have self-tuning or auto-tuning PID control algorithms. Sample times for the microprocessor-based SLCs vary from 0.1 to 0.5 s. Low-pass analog electronic filters are usually installed on the process inputs to minimize aliasing errors caused by high-frequency content in the process signal. Input filter time constants are typically in the range of 0.1 to 1 s. Microprocessor-based SLCs may be made part of a DCS by using the communication port (RS-488 is common) on the controller or may be operated in a stand-alone mode independently of the DCS.

Field Bus Controller Field bus technology is a network-based communications system that interconnects measurement and control equipment such as sensors, actuators, and controllers. Advanced field bus systems, intended for process control applications, such as Foundation Fieldbus, enable digital interoperability among these devices and have a built-in capability to distribute the control application across the network. Several manufacturers have made available Foundation Fieldbus devices that support process controller functionality. These controllers, known as field bus controllers, typically reside in the final control element or measurement transmitter, but can be designed into any field bus device. A suitable communications interface connects the Fieldbus segment to the distributed control system. When configuring the control strategy, all or part of the strategy may be loaded into the field bus devices. The remaining part of the control strategy would reside in the DCS itself. The distribution of the control function depends on the processing capacity of the field bus devices, the control strategy, and where it makes sense to perform these functions. Linearization of a control valve could be performed in the digital valve positioner (controller), for example. Temperature and pressure compensation of a flow measurement could be performed in the flow transmitter processor. The capability of field bus devices varies greatly. Some devices will allow instances of control system function blocks to be loaded and executed, while other devices allow the use of only preconfigured function blocks. Field bus controllers are typically configured as single-loop PID controllers, but cascade or other complex control strategies can be configured depending on the capability of the field bus device. Field bus devices that have native support for process control functions do not necessarily implement the PID algorithm in the same way. It is important to understand these differences so that the controller tuning will deliver the desired closed-loop characteristics. The functionality of field bus devices is projected to increase as the controller market develops.

Controller Reliability and Application Trends Critical process control applications demand a high level of reliability from the electronic controller. Some methods that improve the reliability of electronic controllers include (1) focusing on robust circuit design using quality components; (2) using redundant circuits, modules, or subsystems where necessary; (3) using small backup systems when needed; (4) reducing repair time and using more powerful diagnostics; and (5) distributing functionality to more independent modules to limit the impact of a failed module. Currently, the trend in process control is away from centralized process control and toward an increased number of small distributed control or PLC systems. This trend will put emphasis on the evolution of the field bus controller and continued growth of the PC-based controller. Also, as hardware and software improve, the functionality of the controller will increase, and the supporting hardware will be physically smaller. Hence, the traditional lines between the DCS and the PLC will become less distinct as systems become capable of supporting either function set.

Controller Performance and Process Dynamics The design of a control loop must take the control objectives into account. What do you want this loop to do? And under what operating conditions? There may be control applications that require a single control objective and others that have several control objectives. Control objectives may include such requirements as minimum variance control at steady state, maximum speed of recovery from a major disturbance, maximum speed of set-point response where overshoot and ringing are acceptable, critically damped set-point response with no overshoot, robustness issues, and special start-up or shut-down requirements. The control objectives will define not only the tuning requirements of the controller, but also, to a large extent, the allowable dynamic parameters of the field instruments and process design. Process dynamics alone can prevent control objectives from being realized. Tuning of the controller cannot compensate for an incompatible process or unrealistic control objectives. For most controllers, the difference between the set-point and process feedback signal, the error, is the input to the PID algorithm. The calculated PID output is sent back to the final control element. Every component between the controller output and the process feedback is considered by the controller as the "process" and will directly affect the dynamics and ultimately the performance of the system. This includes not only the dynamics of the physical process, but also the dynamics of the field instruments, signal conditioning equipment, and controller signal processing elements such as filters, scaling, and linearization routines. The choice of final control element can significantly affect the dynamics of the system. If the process dynamics are relatively slow, with time constants of a few minutes or longer, most control valves are fast enough that their contribution to the overall process time response will be negligible. In cases where the process time constants are only a few seconds, the control valve dynamics may become the dominant lag in the overall response. Excessive filtering in the field-sensing devices may also mask the true process dynamics and potentially limit control performance. Often, the design of a control loop and the tuning of the controller are a compromise between a number of different control objectives. When a compromise is unacceptable, gain scheduling or other adaptive tuning routine may be necessary to match the controller response to the most appropriate control objective.

When one is tuning a controller, the form of the PID algorithm must be known. The three common forms of the PID algorithm are parallel or noninteracting, classical or interacting, and the ISA Standard form. In *most* cases, a controller with any of these PID forms can be tuned to produce the desired closed-loop response. The actual tuning parameters will be different. The units of the tuning parameters also affect their value. The controller gain parameter is typically represented as a pure gain (K_c), acting on the error, or as proportional band (PB). In cases where the proportional band parameter is used, the equivalent controller gain is equal to 100 divided by the proportional band and represents the percent span that the error must traverse to produce a 100 percent change in the controller output. The proportional band is always applied to the controller error in terms of percent of span and percent of output. Controllers that use a gain tuning parameter commonly scale the error into percent span and use a percent output basis. In some controllers, the error is scaled by using a separate parameter into percent span prior to the PID algorithm. The gain parameter can also be applied to the error in engineering units. Even though most controller outputs are scaled as a percent, in cascade strategies the controller output may need to be scaled to the same span at the slave loop set point. In this case, the controller gain may in fact be required to calculate the controller output in terms of the slave loop engineering units.

The execution rate of a digital controller should be sufficiently fast, compared to the process dynamics, to produce a response that closely approximates that of an analog controller with the same tuning. A general rule of thumb is that the execution interval should be at least 3 times faster than the dominant lag of the process or about 10 times faster than the closed-loop time constant. The controller can be used when the sample rates are slower than this recommendation, but the controller output will consist of a series of coarse steps as compared to a smooth response. This may create stability problems. Some integral and derivative algorithms may not be accurate when the time-based tuning parameters approach the controller execution interval. The analog inputs of the controller are typically protected from aliasing errors through the use of one- or two-pole analog filters. Faster sample rates allow a smaller antialiasing filter and improved input response characteristics. Some controllers or I/O subsystems oversample the analog inputs with respect to the controller execution interval and then process the input data through a digital filter. This technique can produce a more representative measurement with less quantization noise.

Differences in the PID algorithm, controller parameters, units, and other fundamental control functions highlight the importance of understanding the structure of the controller and the requirement of sufficiently detailed documentation. This is especially important for the controller but is also important for the field instruments, final control elements, and device that have the potential to affect the signal characteristics.

CONTROL VALVES

A control valve consists of a valve, an actuator, and usually one or more valve control devices. The valves discussed in this section are applicable to throttling control (i.e., where flow through the valve is regulated to any desired amount between maximum and minimum limits). Other valves such as check, isolation, and relief valves are addressed in the next subsection. As defined, control valves are automatic control devices that modify the fluid flow rate as specified by the controller.

Valve Types Types of valves are categorized according to their design style. These styles can be grouped into type of stem motion—linear or rotary. The valve stem is the rod, shaft, or spindle that connects the actuator with the closure member (i.e., a movable part of the valve that is positioned in the flow path to modify the rate of flow). Movement of either type of stem is known as travel. The major categories are described briefly below.

Globe and Angle The most common linear stem-motion control valve is the globe valve. The name comes from the globular cavities around the port. In general, a port is any fluid passageway, but often the reference is to the passage that is blocked off by the closure member when the valve is closed. In globe valves, the closure member is called a plug. The plug in the valve shown in Fig. 8-66 is guided by a large-diameter port and moves within the port to provide the flow control orifice of the valve. A very popular alternate construction is a cage-guided plug, as illustrated in Fig. 8-67. In many such designs,

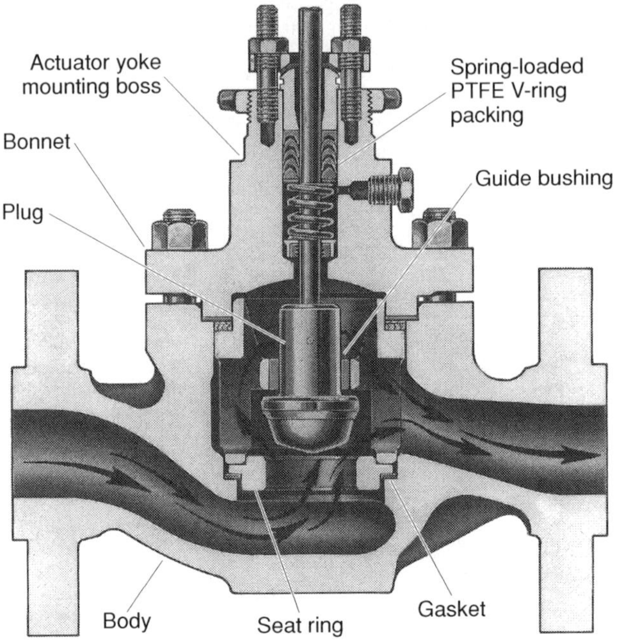

FIG. 8-66 Post-guided contour plug globe valve with metal seat and raised-face flange end connections. (*Courtesy Fisher Controls International LLC.*)

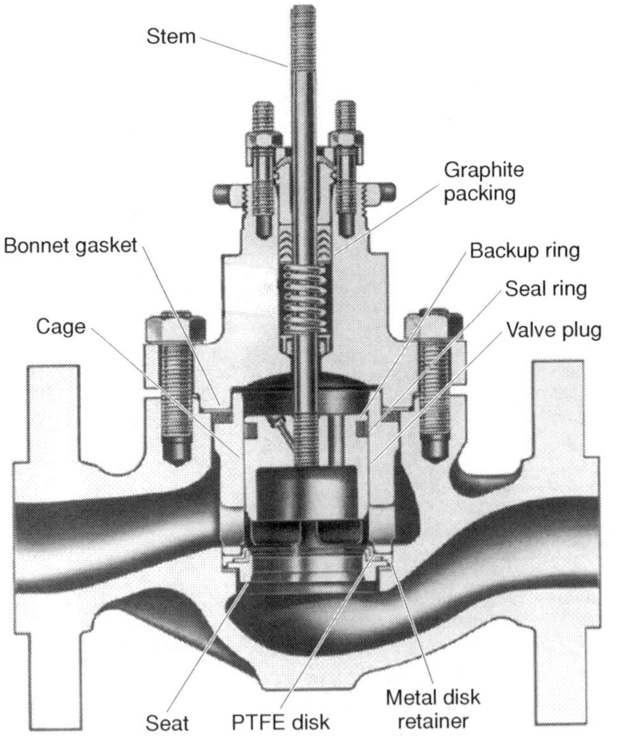

FIG. 8-67 Cage-guided balanced plug globe valve with polymer seat and plug seal. (*Courtesy Fisher Controls International LLC.*)

openings in the cage provide the flow control orifices. The valve seat is the zone of contact between the moving closure member and the stationary valve body, which shuts off the flow when the valve is closed. Often the seat in the body is on a replaceable part known as a seat ring. This stationary seat can also be designed as an integral part of the cage. Plugs may also be port-guided by wings or a skirt that fits snugly into the seat-ring bore.

One distinct advantage of cage guiding is the use of balanced plugs in single-port designs. The unbalanced plug depicted in Fig. 8-66 is subjected to a static pressure force equal to the port area times the valve pressure differential (plus the stem area times the downstream pressure) when the valve is closed. In the balanced design (Fig. 8-67), note that both the top and bottom of the plug are subjected to the same downstream pressure when the valve is closed. Leakage via the plug-to-cage clearance is prevented by a plug seal. Both plug types are subjected to hydrostatic force due to internal pressure acting on the stem area and to dynamic flow forces when the valve is flowing.

The plug, cage, seat ring, and associated seals are known as the trim. A key feature of globe valves is that they allow maintenance of the trim via a removable bonnet without removing the valve body from the line. Bonnets are typically bolted on but may be threaded in smaller sizes.

Angle valves are an alternate form of the globe valve. They often share the same trim options and have the top-entry bonnet style. Angle valves can eliminate the need for an elbow but are especially useful when direct impingement of the process fluid on the body wall is to be avoided. Sometimes it is not practical to package a long trim within a globe body, so an angle body is used. Some angle bodies are self-draining, which is an important feature for dangerous fluids.

Butterfly The classic design of butterfly valves is shown in Fig. 8-68. Its chief advantage is high capacity in a small package and a very low initial cost. Much of the size and cost advantage is due to the wafer body design, which is clamped between two pipeline flanges. In the simplest design, there is no seal as such, merely a small clearance gap between the disc OD and the body ID. Often a true seal is provided by a resilient material in the body that is engaged via an interference fit with the disc. In a lined butterfly valve, this material covers the entire body ID and extends around the body ends to eliminate the need for pipeline joint gaskets. In a fully lined valve, the disc is also coated to minimize corrosion or erosion.

A high-performance butterfly valve has a disc that is offset from the shaft centerline. This eccentricity causes the seating surface to move away from the seal once the disc is out of the closed position, reducing friction and seal wear. It is also known as an eccentric disc valve; the advantage of the butterfly valve is improved shutoff while maintaining high ultimate capacity at a reasonable cost. This cost advantage relative to other design styles is particularly true in sizes above 6-in nominal pipe size (NPS). Improved shutoff is due to advances in seal technologies, including polymer, flexing metal, combination metal with polymer inserts, and so on, many utilizing pressure assist.

Ball Ball valves get their name from the shape of the closure member. One version uses a full spherical member with a cylindrical bore through it. The ball is rotated one-quarter turn from the full-closed to the full-open position. If the bore is the same diameter as the mating-pipe fitting ID, the valve is referred to as full-bore. If the hole is undersized, the ball valve is considered to be a venturi style. A segmented ball is a portion of a hollow sphere that is large enough to block the port when closed. Segmented balls often have a V-shaped contour along one edge, which provides a desirable flow characteristic (see Fig. 8-69). Both full ball and segmented ball valves are known for

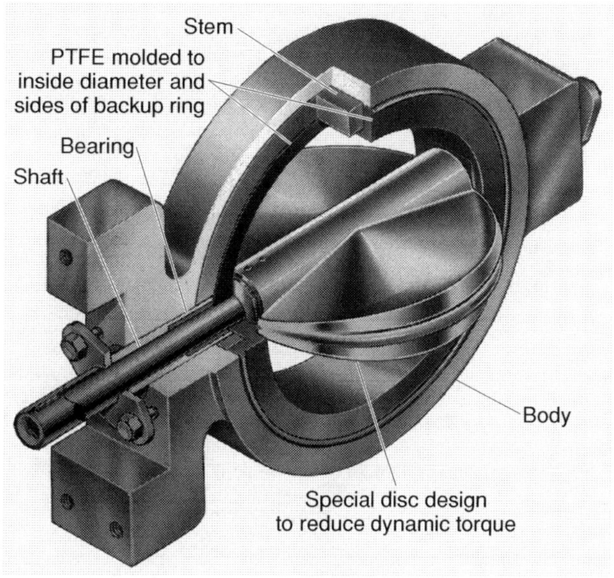

FIG. 8-68 Partial cutaway of wafer-style lined butterfly valve. (*Courtesy Fisher Controls International LLC.*)

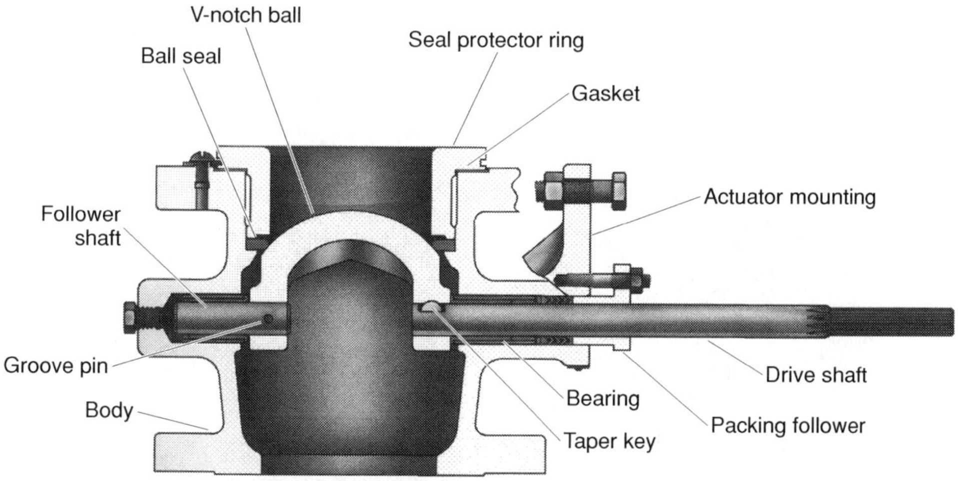

FIG. 8-69 Segmented ball valve. Partial view of actuator mounting shown 90° out of position. (*Courtesy Fisher Controls International LLC.*)

their low resistance to flow when full open. Shutoff leakage is minimized through the use of flexing or spring-loaded elastomeric or metal seals. Bodies are usually in two or three pieces or have a removable retainer to facilitate installing seals. End connections are usually flanged or threaded in small sizes, although segmented ball valves are offered in wafer style also.

Plug There are two substantially different rotary valve design categories referred to as plug valves. The first consists of a cylindrical or slightly conical plug with a port through it. The plug rotates to vary the flow much as a ball valve does. The body is top-entry but is geometrically simpler than a globe valve and thus can be lined with fluorocarbon polymer to protect against corrosion. These plug valves have excellent shutoff but are generally not for modulating service due to high friction. A variation of the basic design (similar to the eccentric butterfly disc) only makes sealing contact in the closed position and is used for control.

The other rotary plug design is portrayed in Fig. 8-70. The seating surface is substantially offset from the shaft, producing a ball-valve-like motion with the additional cam action of the plug into the seat when closing. In reverse flow, high-velocity fluid motion is directed inward, impinging on itself and only contacting the plug and seat ring.

Multiport This term refers to any valve or manifold of valves with more than one inlet or outlet. For throttling control, the three-way body is used for blending (two inlets, one outlet) or as a diverter (one inlet, two outlets). A three-way valve is most commonly a special globelike body with special trim that allows flow both over and under the plug. Two rotary valves and a pipe tee can also be used. Special three-, four-, and five-way ball valve designs are used for switching applications.

Special Application Valves

Digital Valves True digital valves consist of discrete solenoid-operated flow ports that are sized according to binary weighing. The valve can be designed with sharp-edged orifices or with streamlined nozzles that can be used for flow metering. Precise control of the throttling control orifice is the strength of the digital valve. Digital valves are mechanically complicated and expensive, and they have considerably reduced maximum flow capacities compared to the globe and rotary valve styles.

Cryogenic Service Valves designed to minimize heat absorption for throttling liquids and gases below 80 K are called cryogenic service valves. These valves are designed with small valve bodies to minimize heat absorption and long bonnets between the valve and actuator to allow for extra layers of insulation around the valve. For extreme cases, vacuum jacketing can be constructed around the entire valve to minimize heat influx.

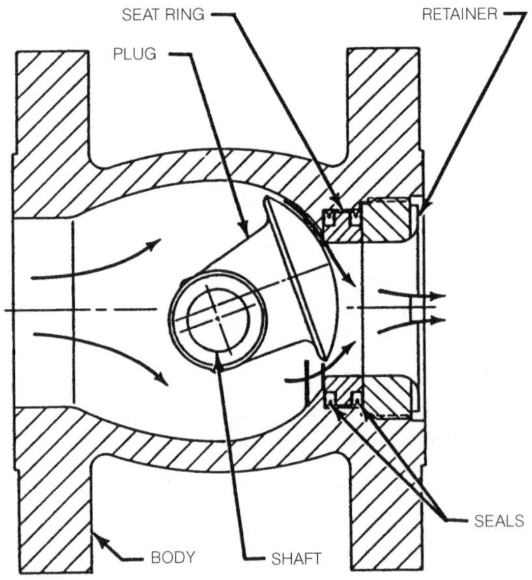

FIG. 8-70 Eccentric plug valve shown in erosion-resistant reverse flow direction. Shaded components can be made of hard metal or ceramic materials. (*Courtesy Fisher Controls International LLC.*)

High Pressure Valves used for pressures nominally above 760 bar (11,000 psi, pressures above ANSI Class 4500) are often custom-designed for specific applications. Normally, these valves are of the plug type and use specially hardened plug and seat assemblies. Internal surfaces are polished, and internal corners and intersecting bores are smoothed to reduce high localized stresses in the valve body. Steam loops in the valve body are available to raise the body temperature to increase ductility and impact strength of the body material.

High-Viscous Process Used most extensively by the polymer industry, the valve for high-viscous fluids is designed with smooth finished internal passages to prevent stagnation and polymer degradation. These valves are available with integral body passages through which a heat-transfer fluid is pumped to keep the valve and process fluid heated.

Pinch The industrial equivalent of controlling flow by pinching a soda straw is the pinch valve. Valves of this type use fabric-reinforced elastomer sleeves that completely isolate the process fluid from the metal parts in the valve. The valve is actuated by applying air pressure directly to the outside of the sleeve, causing it to contract or pinch. Another method is to pinch the sleeve with a linear actuator with a specially attached foot. Pinch valves are used extensively for corrosive material service and erosive slurry service. This type of valve is used in applications with pressure drops up to 10 bar (145 psi).

Fire-Rated Valves that handle flammable fluids may have additional safety-related requirements for minimal external leakage, minimal internal (downstream) leakage, and operability during and after a fire. Being fire-rated does not mean being totally impervious to fire, but a sample valve must meet particular specifications such as those of American Petroleum Institute (API) 607, Factory Mutual Research Corp. (FM) 7440, or the British Standard 5146 under a simulated fire test. Due to very high flame temperature, metal seating (either primary or as a backup to a burned-out elastomer) is mandatory.

Solids Metering The control valves described earlier are primarily used for the control of fluid (liquid or gas) flow. Sometimes these valves, particularly the ball, butterfly, or sliding gate valves, are used to throttle dry or slurry solids. More often, special throttling mechanisms such as venturi ejectors, conveyers, knife-type gate valves, or rotating vane valves are used. The particular solids-metering valve hardware depends on the volume, density, particle shape, and coarseness of the solids to be handled.

Actuators An actuator is a device that applies the force (torque) necessary to cause a valve's closure member to move. Actuators must overcome pressure and flow forces as well as friction from packing, bearings or guide surfaces, and seals; and must provide the seating force. In rotary valves, maximum friction occurs in the closed position, and the moment necessary to overcome it is referred to as breakout torque. The rotary valve shaft torque generated by steady-state flow and pressure forces is called dynamic torque. It may tend to open or close the valve depending on valve design and travel. Dynamic torque per unit pressure differential is largest in butterfly valves at roughly 70° open. In linear stem-motion valves, the flow forces should not exceed the available actuator force, but this is usually accounted for by default when the seating force is provided.

Actuators often provide a fail-safe function. In the event of an interruption in the power source, the actuator will place the valve in a predetermined safe position, usually either full-open or full-closed. Safety systems are often designed to trigger local fail-safe action at specific valves to cause a needed action to occur, which may not be a complete process or plant shutdown.

Actuators are classified according to their power source. The nature of these sources leads naturally to design features that make their performance characteristics distinct.

Pneumatic Despite the availability of more sophisticated alternatives, the pneumatically driven actuator is still by far the most popular type. Historically the most common has been the spring and diaphragm design (Fig. 8-71). The compressed air input signal fills a

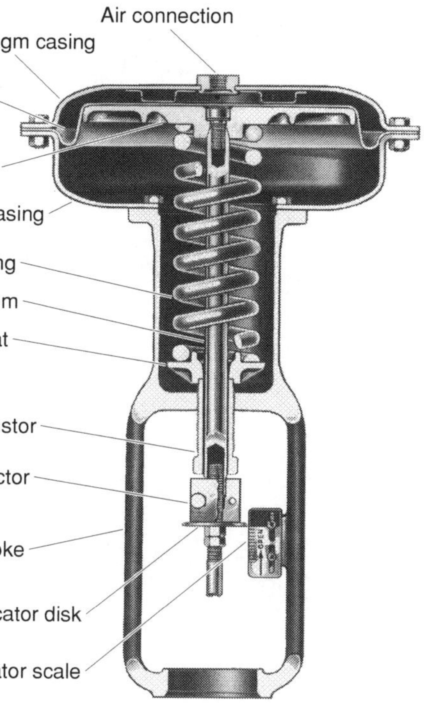

FIG. 8-71 Spring and diaphragm actuator with an "up" fail-safe mode. Spring adjuster allows slight alteration of bench set. (*Courtesy Fisher Controls International LLC.*)

chamber sealed by an elastomeric diaphragm. The pressure force on the diaphragm plate causes a spring to be compressed and the actuator stem to move. This spring provides the fail-safe function and contributes to the dynamic stiffness of the actuator. If the accompanying valve is "push down to close," the actuator depicted in Fig. 8-71 will be described as "air to close" or synonymously as fail-open. A slightly different design yields "air to open" or fail-closed action. The spring is typically precompressed to provide a significant available force in the failed position (e.g., to provide seating load). The spring also provides a proportional relationship between the force generated by air pressure and stem position. The pressure range over which a spring and diaphragm actuator strokes in the absence of valve forces is known as the bench set. The chief advantages of spring and diaphragm actuators are their high reliability, low cost, adequate dynamic response, and fail-safe action—all of which are inherent in their simple design.

Alternately, the pressurized chamber can be formed by a circular piston with a seal on its outer edge sliding within a cylindrical bore. Higher operating pressure [6 bar (~ 90 psig) is typical] and longer strokes are possible. Piston actuators can be spring-opposed but many times are in a dual-acting configuration (i.e., compressed air is applied to both sides of the piston with the net force determined from the pressure difference—see Fig. 8-72). Dynamic stiffness is usually higher with piston designs than with spring and diaphragm actuators; see "Positioner/Actuator Stiffness." Fail-safe action, if necessary, is achieved without a spring through the use of additional solenoid valves, trip valves, or relays. See "Valve Control Devices."

Motion Conversion Actuator power units with translational output can be adapted to rotary valves that generally need 90° or less rotation. A lever is attached to the rotating shaft, and a link with pivoting means on the end connects to the linear output of the power unit, an arrangement similar to an internal combustion engine crankshaft, connecting rod, and piston. When the actuator piston, or more commonly the diaphragm plate, is designed to tilt, one pivot can be eliminated (see Fig. 8-72). Scotch yoke and rack-and-pinion arrangements are also commonly used, especially with piston power units.

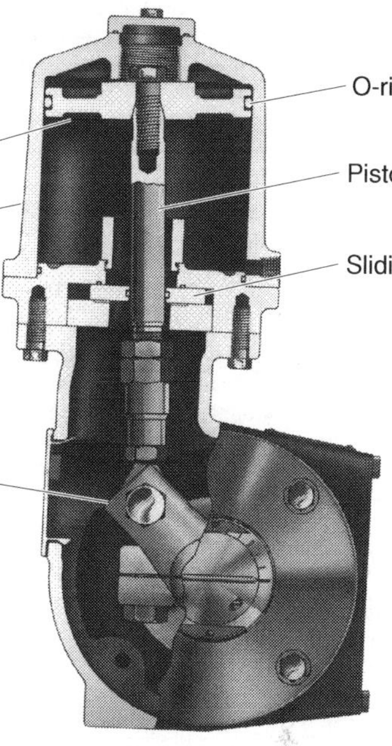

FIG. 8-72 Double-acting piston rotary actuator with lever and tilting piston for motion conversion. (*Courtesy Fisher Controls International LLC.*)

Friction and the changing mechanical advantage of these motion conversion mechanisms mean the available torque may vary greatly with travel. One notable exception is vane-style rotary actuators whose offset "piston" pivots, giving direct rotary output.

Hydraulic The design of typical hydraulic actuators is similar to that of double-acting piston pneumatic types. One key advantage is the high pressure [typically 35 to 70 bar (500 to 1000 psi)], which leads to high thrust in a smaller package. The incompressible nature of the hydraulic oil means these actuators have very high dynamic stiffness. The incompressibility and small chamber size connote fast stroking speed and good frequency response. The disadvantages include high initial cost, especially when considering the hydraulic supply. Maintenance is much more difficult than with pneumatics, especially on the hydraulic positioner.

Electrohydraulic actuators have similar performance characteristics and cost/maintenance ramifications. The main difference is that they contain their own electric-powered hydraulic pump. The pump may run continuously or be switched on when a change in position is required. Their main application is remote sites without an air supply when a fail-safe spring return is needed.

Electric The most common electric actuators use a typical motor—three-phase ac induction, capacitor-start split-phase induction, or dc. Normally the motor output passes through a large gear reduction and, if linear motion output is required, a ball screw or thread. These devices can provide large thrust, especially given their size. Lost motion in the gearing system does create backlash, but if not operating across a thrust reversal, this type of actuator has very high stiffness. Usually the gearing system is self-locking, which means that forces on the closure member cannot move it by spinning a nonenergized motor. This behavior is called a lock-in-last-position fail-safe mode. Some gear systems (e.g., low-reduction spur gears) can be backdriven. A solenoid-activated mechanical brake or locking current to motor field coils is added to provide lock-in-last-position fail-safe mode. A battery backup system for a dc motor can guard against power failures. Otherwise, an electric actuator is not acceptable if fail-open/closed action is mandatory. Using electric power requires environmental enclosures and explosion protection, especially in hydrocarbon processing facilities; see the full discussion in "Valve Control Devices."

Unless sophisticated speed control power electronics is used, position modulation is achieved via bang-zero-bang control. Mechanical inertia causes overshoot, which is (1) minimized by braking and/or (2) hidden by adding dead band to the position control. Without these provisions, high starting currents would cause motors to overheat from constant "hunting" within the position loop. Travel is limited with power interruption switches or with force (torque) electromechanical cutouts when the closed position is against a mechanical stop (e.g., a globe valve). Electric actuators are often used for on/off service. Stepper motors can be used instead, and they, as their name implies, move in fixed incremental steps. Through gear reduction, the typical number of increments for 90° rotation ranges from 5000 to 10,000; hence positioning resolution at the actuator is excellent.

An electromagnetic solenoid can be used to directly actuate the plug on very small linear stem-motion valves. A solenoid is usually designed as a two-position device, so this valve control is on/off. Special solenoids with position feedback can provide proportional action for modulating control. Force requirements of medium-sized valves can be met with piloted plug designs, which use process pressure to assist the solenoid force. Piloted plugs are also used to minimize the size of common pneumatic actuators, especially when there is need for high seating load.

Manual A manually positioned valve is by definition not an automatic control valve, but it may be involved with process control. For rotary valves, the manual operator can be as simple as a lever, but a wheel driving a gear reduction is necessary in larger valves. Linear motion is normally created with a wheel turning a screw-type device. A manual override is usually available as an option for the powered actuators listed above. For spring-opposed designs, an adjustable travel stop will work as a one-way manual input. In more complex designs, the handwheel can provide loop control override via an engagement means. Some gear reduction systems of electric actuators allow the manual positioning to be independent of the automatic positioning without declutching.

OTHER PROCESS VALVES

In addition to the throttling control valve, other types of process valves can be used to manipulate a process.

Valves for On/Off Applications Valves are often required for service that is primarily nonthrottling. Valves in this category, depending on the service requirements, may be of the same design as the types used for throttling control or, as in the case of gate valves, different in design. Valves in this category usually have tight shutoff when they are closed and low pressure drops when they are wide open. The on/off valve can be operated manually, such as by handwheel or lever; or automatically, with pneumatic or electric actuators.

Batch Batch process operation is an application requiring on/off valve service. Here the valve is opened and closed to provide reactant, catalyst, or product to and from the batch reactor. Like the throttling control valve, the valve used in this service must be designed to open and close thousands of times. For this reason, valves used in this application are often the same valves used in continuous throttling applications. Ball valves are especially useful in batch operations. The ball valve has a straight-through flow passage that reduces pressure drop in the wide-open state and provides tight shutoff capability when closed. In addition, the segmented ball valve provides for shearing action between the ball and the ball seat that promotes closure in slurry service.

Isolation A means for pressure-isolating control valves, pumps, and other piping hardware for installation and maintenance is another common application for an on/off valve. In this application, the valve is required to have tight shutoff so that leakage is stopped when the piping system is under repair. As the need to cycle the valve in this application is far less than that of a throttling control valve, the wear characteristics of the valve are less important. Also, because many are required in a plant, the isolation valve needs to be reliable, simple in design, and simple in operation.

The gate valve, shown in Fig. 8-73, is the most widely used valve in this application. The gate valve is composed of a gatelike disc that moves perpendicular to the flow stream. The disc is moved up and down by a threaded screw that is rotated to effect disc movement. Because the disc is large and at right angles to the process pressure, a large seat loading for tight shutoff is possible. Wear produced by high seat loading during the movement of the disc prohibits the use of the gate valve for throttling applications.

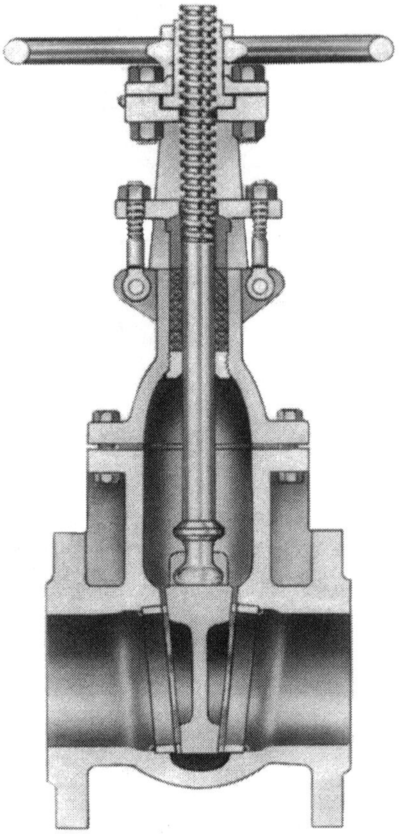

FIG. 8-73 Gate valve. (*Courtesy Crane Valves.*)

Pressure Relief Valves The pressure relief valve is an automatic pressure-relieving device designed to open when normal conditions are exceeded and to close again when normal conditions are restored. Within this class there are relief valves, pilot-operated pressure relief valves, and safety valves. Relief valves (see Fig. 8-74) have spring-loaded discs that close a main orifice against a pressure source. As pressure rises, the disc begins to rise off the orifice and a small amount of fluid passes through the valve. Continued rise in pressure above the opening pressure causes the disc to open the orifice in a proportional fashion. The main orifice reduces and closes when the pressure returns to the set pressure. Additional sensitivity to overpressure conditions can be improved by adding an auxiliary pressure relief valve (pilot) to the basic pressure relief valve. This combination is known as a pilot-operated pressure relief valve.

The safety valve is a pressure relief valve that is designed to open fully, or pop, with only a small amount of pressure over the rated limit. Where conventional safety valves are sensitive to downstream pressure and may have unsatisfactory operating characteristics in variable backpressure applications, pressure-balanced safety relief valve designs are available to minimize the effect of downstream pressure on performance.

Application and sizing of pressure relief valves, pilot-operated pressure relief valves, and safety valves for use on pressure vessels are found in the ASME Boiler and Pressure Vessel Code, Section VIII, Division 1, "Rules for Construction of Pressure Vessels," Paragraphs UG-125 through UG-137.

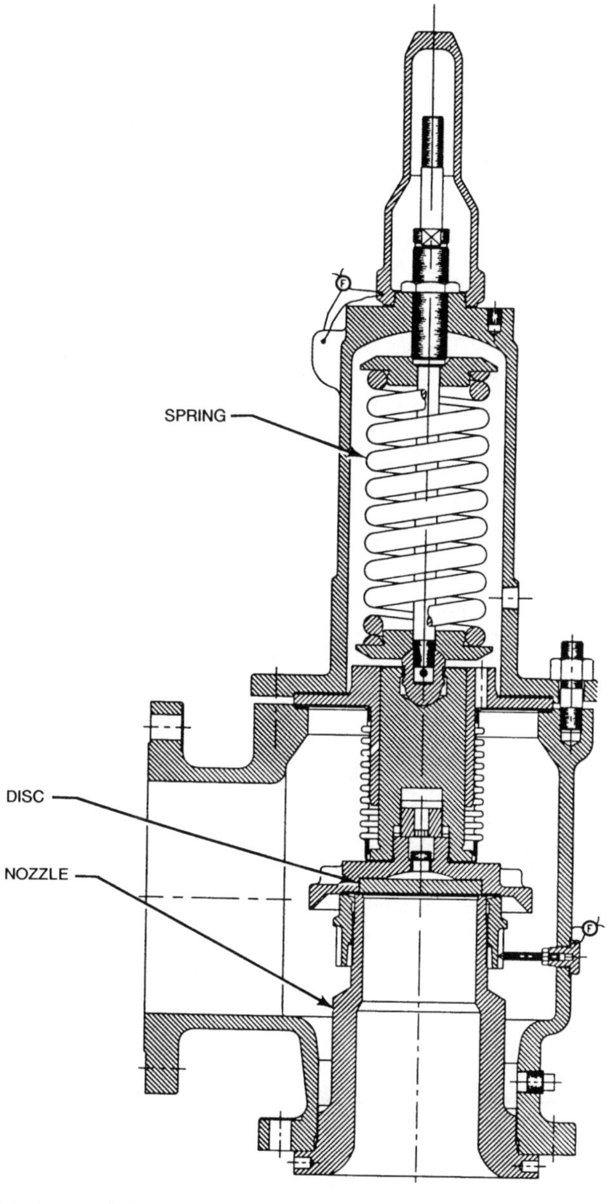

FIG. 8-74 Relief valve. (*Courtesy of Teledyne Fluid Systems, Farris Engineering.*)

Check Valves The purpose of a check valve is to allow relatively unimpeded flow in the desired direction but to prevent flow in the reverse direction. Two common designs are swing-type and lift-type check valves, the names of which denote the motion of the closure member. In the forward direction, flow forces overcome the weight of the member or a spring to open the flow passage. With reverse pressure conditions, flow forces drive the closure member into the valve seat, thus providing shutoff.

VALVE DESIGN CONSIDERATIONS

Functional requirements and the properties of the controlled fluid determine which valve and actuator types are best for a specific application. If demands are modest and no unique valve features are required, the valve design style selection may be determined solely by cost. If so, general-purpose globe or angle valves provide exceptional value, especially in sizes less than 3-in NPS and hence are very popular. Beyond type selection, there are many other valve specifications that must be determined properly to ultimately yield improved process control.

Materials and Pressure Ratings Valves must be constructed from materials that are sufficiently immune to corrosive or erosive action by the process fluid. Common body materials are cast iron, steel, stainless steel, high-nickel alloys, and copper alloys such as bronze. Trim materials need better corrosion and erosion resistance due to the higher fluid velocity in the throttling region. High hardness is desirable in erosive and cavitating applications. Heat-treated and precipitation-hardened stainless steels are common. High hardness is also good for guiding, bearing, and seating surfaces; cobalt-chromium alloys are utilized in cast or wrought form and frequently as welded overlays called hard facing. In less stringent situations, chrome plating, heat-treated nickel coatings, and nitriding are used. Tungsten carbide and ceramic trim are warranted in extremely erosive services. See Sec. 25, "Materials of Construction," for specific material properties.

Since the valve body is a pressurized vessel, it is usually designed to comply with a standardized system of pressure ratings. Two common systems are described in the standards ASME B16.34 and EN 12516. Internal pressure limits under these standards are divided into broad classes, with specific limits being a function of material and temperature. Manufacturers also assign their own pressure ratings based on internal design rules. A common insignia is *250 WOG*, which means a pressure rating of 250 psig (~17 bar) in water, oil, or gas at ambient temperature. "Storage and Process Vessels" in Sec. 10 provides introductory information on compliance of pressure vessel design to industry codes (e.g., ASME Boiler and Pressure Vessel Code, Section VIII; ASME B31.3 Chemical Plant and Petroleum Refinery Piping).

Valve bodies are also standardized to mate with common piping connections: flanged, butt-welded end, socket-welded end, and screwed end. Dimensional information for some of these joints and class pressure-temperature ratings are included in Sec. 10, "Process Plant Piping." Control valves have their own standardized face-to-face dimensions that are governed by ANSI/ISA Standards S75.08 and S75.22. Butterfly valves are also governed by API 609 and Manufacturers Standardization Society (MSS) SP-67 and SP-68.

Sizing Throttling control valves must be selected to pass the required flow rate, given expected pressure conditions. Sizing is not merely matching the end connection size with surrounding piping; it is a key step in ensuring that the process can be properly controlled. Sizing methods range from simple models based on elementary fluid mechanics to very complex models when unusual thermodynamics or nonideal behaviors occur. Basic sizing practices have been standardized (for example, ANSI-75.01.01) and are implemented as PC-based programs by manufacturers. The following is a discussion of very basic sizing equations and the associated physics.

Regardless of the particular process variable being controlled (e.g., temperature, level, pH), the output of a control valve is the flow rate. The throttling valve performs its function of manipulating the flow rate by virtue of being an adjustable resistance to flow. Flow rate and pressure conditions are normally known when a process is designed, and the valve resistance range must be matched accordingly. In the tradition of orifice and nozzle discharge coefficients, this resistance is embodied in the valve flow coefficient C_V. By applying the principles of conversation of mass and energy, the mass flow rate w kg/h is given for a liquid by

$$w = 27.3C_V\sqrt{\rho(p_1 - p_2)} \tag{8-111}$$

where p_1 and p_2 are upstream and downstream static pressure, bar, respectively. The density of the fluid ρ is expressed in kilograms per cubic meter. This equation is valid for nonvaporizing, turbulent flow conditions for a valve with no attached fittings.

While the above equation gives the relationship between pressure and flow from a macroscopic point of view, it does not explain what is going on inside the valve. Valves create a resistance to flow by restricting the cross-sectional area of the flow passage and by forcing the fluid to change direction as it passes through the body and trim.

The conservation of mass principle dictates that, for steady flow, the product of density, average velocity, and cross-sectional area remain a constant. The average velocity of the fluid stream at the minimum restriction in the valve is therefore much higher than that at the inlet. Note that due to the abrupt nature of the flow contraction that forms the minimum passage, the main fluid stream may separate from the passage walls and form a jet that has an even smaller cross section, the so-called vena contracta. The ratio of minimum stream area to the corresponding passage area is called the contraction coefficient. As the fluid expands from the minimum cross-sectional area to the full passage area in the downstream piping, large amounts of turbulence are generated. Direction changes can also induce significant amounts of turbulence.

Figure 8-75 is an illustration of how the mean pressure changes as fluid moves through a valve. Some of the potential energy that was stored in the fluid by pressurizing it (e.g., the work done by a pump) is first converted to the kinetic energy of the fast-moving fluid at the vena contracta. Some of that kinetic energy turns into the kinetic energy of turbulence. As the turbulent eddies break down into smaller and smaller structures, viscous effects ultimately convert all the turbulent energy to heat. Therefore, a valve converts fluid energy from one form to another.

For many valve constructions, it is reasonable to approximate the fluid transition from the valve inlet to the minimum cross section of the flow stream as an isentropic or lossless process. Using this approximation, the minimum pressure p_{VC} can be estimated from the Bernoulli relationship. See Sec. 6 ("Fluid and Particle Mechanics") for more background information. Downstream of the vena contracta, the flow is definitely not lossless due to all the turbulence that is generated. As the flow passage area increases and the fluid slows down, some of the kinetic energy of the fluid is converted back to pressure energy as pressure recovers. The energy that is permanently lost via turbulence accounts for the permanent pressure or head loss of the valve. The relative amount of pressure that is recouped determines whether the valve is considered to be high- or low-recovery. The flow passage geometry at and downstream of the vena contracta primarily determines the amount of recovery. The amount of recovery is quantified by the liquid pressure recovery factor F_L

$$F_L = \sqrt{\frac{p_1 - p_2}{p_1 - p_{VC}}} \qquad (8\text{-}112)$$

Under some operating conditions, sufficient pressure differential may exist across the valve to cause the vena contracta pressure to drop to the saturation pressure (also known as the *vapor pressure*) of the liquid. If this occurs, a portion of the liquid will vaporize, forming a two-phase, compressible mixture within the valve. If sufficient vapor forms, the flow may *choke*. When a flow is choked, any increase in pressure differential across the valve no longer produces an increase in flow through the valve.

The vena contracta condition at choked flow for pure liquids has been shown to be

$$p_{VC} = F_F p_v \qquad (8\text{-}113)$$

where

$$F_F = 0.96 - 0.28\sqrt{\frac{p_v}{p_c}} \qquad (8\text{-}114)$$

and p_{VC} is the absolute vena contracta pressure under choked conditions, F_F is the liquid critical pressure ratio factor, p_v is the absolute vapor pressure of the liquid at inlet temperature, and p_c is the absolute thermodynamic critical pressure of the fluid.

Equations (8-112) and (8-113) can be used together to determine the pressure differential across the valve at the point where the flow chokes

$$\Delta p_{\text{choked}} = F_L^2(p_1 - F_F p_v) \qquad (8\text{-}115)$$

The pressure recovery factor is a constant for any given valve at a given opening. The value of this factor can be established by flow test and is published by the valve manufacturer. If the actual pressure differential across the valve is greater than the choked pressure differential of Eq. (8-115), then Δp_{choked} should be used in Eq. (8-111) to determine the correct valve size. A more complete presentation of sizing relationships is given in ANSI 75.01.01, including provisions for pipe reducers and Reynolds number effects.

Equations (8-111) to (8-115) are restricted to incompressible fluids. For gases and vapors, the fluid density is dependent on pressure. For convenience, compressible fluids are often assumed to follow the ideal gas law model. Deviations from ideal behavior are corrected for, to first order, with nonunity values of compressibility factor Z (see Sec. 2, "Physical and Chemical Data," for definitions and data for common fluids). For compressible fluids

$$w = 94.8 C_V p_1 Y \sqrt{\frac{x M_\omega}{T_1 Z}} \qquad (8\text{-}116)$$

where p_1 is in bar absolute, T_1 is inlet temperature in kelvins, M_w is the molecular weight, and x is the dimensionless pressure drop ratio $(p_1 - p_2)/p_1$. The expansion factor Y accounts for changes in the fluid density as the fluid passes through the valve. It is dependent on pressure drop and valve geometry. Experimental data have shown that for

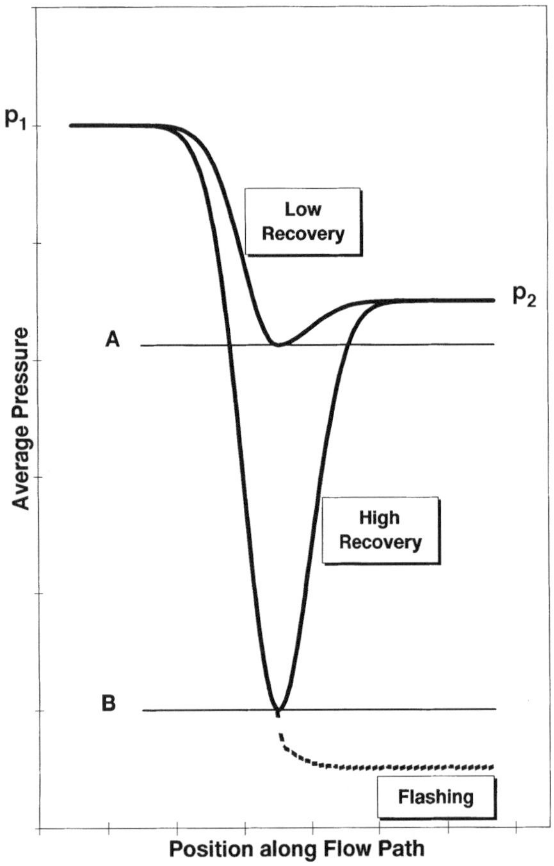

FIG. 8-75 Generic depictions of average pressure at subsequent cross sections throughout a control valve. The F_L values selected for illustration are 0.9 and 0.63 for low and high recovery, respectively. Internal pressure in the high-recovery valve is shown as a dashed line for flashing conditions ($p_2 < p_v$) with $p_v = B$.

small departures in the ratio of specific heat from that of air (1.4), a simple linear relationship can be used to represent the expansion factor:

$$Y = 1 - \frac{1.4x}{3x_T\gamma} \quad \text{for} \quad x \leq \frac{x_T\gamma}{1.4} \tag{8-117}$$

where γ is the ratio of specific heats and x_T is an experimentally determined factor for a specific valve and is the largest value of x that contributes to flow (i.e., values of x greater than x_T do not contribute to flow).

The terminal value of x, x_T, results from a phenomenon known as choking. Given a nozzle geometry with fixed inlet conditions, the mass flow rate will increase as p_2 is decreased up to a maximum amount at the critical pressure drop. The velocity at the vena contracta has reached sonic, and a standing shock has formed. This shock causes a step change in pressure as flow passes through it, and further reduction in p_2 does not increase mass flow. Thus x_T relates to the critical pressure drop ratio and also accounts for valve geometry effects. The value of x_T varies with flow path geometry and is supplied by the valve manufacturer. In the choked case,

$$x > \frac{x_T\gamma}{1.4} \quad \text{and} \quad Y = 0.67 \tag{8-118}$$

Noise Control Sound is a fluctuation of air pressure that can be detected by the human ear. Sound travels through any fluid (e.g., the air) as a compression/expansion wave. This wave travels radially outward in all directions from the sound source. The pressure wave induces an oscillating motion in the transmitting medium that is superimposed on any other net motion it may have. These waves are reflected, refracted, scattered, and absorbed as they encounter solid objects. Sound is transmitted through solids in a complex array of types of elastic waves. Sound is characterized by its amplitude, frequency, phase, and direction of propagation.

Sound strength is therefore location-dependent and is often quantified as a sound pressure level L_p in decibels based on the root mean square (rms) sound pressure value p_S, where

$$L_p = 10 \log\left(\frac{p_S}{p_{\text{reference}}}\right)^2 \tag{8-119}$$

For airborne sound, the reference pressure is 2×10^{-5} Pa (29×10^{-1} psi), which is nominally the human threshold of hearing at 1000 Hz. The corresponding sound pressure level is 0 dB. A voice in conversation is about 50 dB, and a jackhammer operator is subject to 100 dB. Extreme levels such as a jet engine at takeoff might produce 140 dB at a distance of 3 m, which is a pressure amplitude of 200 Pa (29×10^{-3} psi). These examples demonstrate both the sensitivity and the wide dynamic range of the human ear.

Traveling sound waves carry energy. Sound intensity I is a measure of the power passing through a unit area in a specified direction and is related to p_S. Measuring sound intensity in a process plant gives clues to the location of the source. As one moves away from the source, the fact that the energy is spread over a larger area requires that the sound pressure level decrease. For example, doubling one's distance from a point source reduces L_p by 6 dB. Viscous action from the induced fluid motion absorbs additional acoustic energy. However, in free air, this viscous damping is negligible over short distances (on the order of 1 m).

Noise is a group of sounds with many nonharmonic frequency components of varying amplitudes and random phase. The turbulence generated by a throttling valve creates noise. As a valve converts potential energy to heat, some of the energy becomes acoustic energy as an intermediate step. Valves handling large amounts of compressible fluid through a large pressure change create the most noise because more total power is being transformed. Liquid flows are noisy only under special circumstances, as will be seen in the next subsection. Due to the random nature of turbulence and the broad distribution of length and velocity scales of turbulent eddies, valve-generated sound is usually random, broad-spectrum noise. The total sound pressure level from two such statistically uncorrelated sources is (in decibels)

$$L_p = 10 \log \frac{(p_{S1})^2 + (p_{S2})^2}{(p_{\text{reference}})^2} \tag{8-120}$$

For example, two sources of equal strength combine to create an L_p that is 3 dB higher.

While noise is annoying to listen to, the real reasons for being concerned about noise relate to its impact on people and equipment. Hearing loss can occur due to long-term exposure to moderately high, or even short exposure to very high, noise levels. The U.S. Occupational Safety and Health Act (OSHA) has specific guidelines for permissible levels and exposure times. The human ear has a frequency-dependent sensitivity to sound. When the effect on humans is the criterion, L_p measurements are weighted to account for the ear's response. This so-called A-weighted scale is defined in ANSI S1.4 and is commonly reported as L_{pA}. Figure 8-76 illustrates the difference between actual and perceived airborne sound pressure levels. At sufficiently high levels, noise and the associated vibration can damage equipment.

There are two approaches to fluid-generated noise control—source or path treatment. Path treatment means absorbing or blocking the transmission of noise after it has been created. The pipe itself is a barrier. The sound pressure level inside a standard schedule pipe is roughly 40 to 60 dB higher than on the outside. Thicker-walled pipe reduces levels somewhat more, and adding acoustical insulation on the outside of the pipe reduces ambient levels up to 10 dB per inch of thickness. Since noise propagates relatively unimpeded inside the

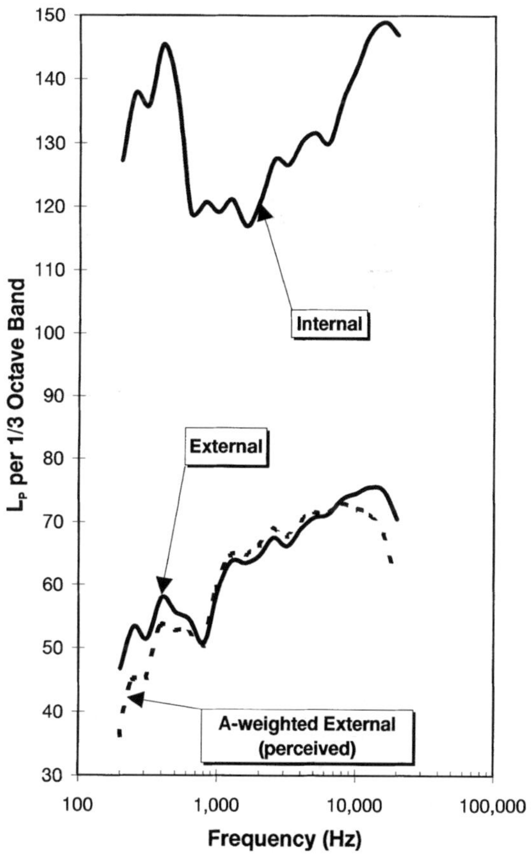

FIG. 8-76 Valve-generated sound pressure level spectrums.

pipe, barrier approaches require the entire downstream piping system to be treated in order to be totally effective. In-line silencers place absorbent material inside the flow stream, thus reducing the level of the internally propagating noise. Noise reductions up to 25 dB can be achieved economically with silencers.

The other approach to valve noise problems is the use of quiet trim. Two basic strategies are used to reduce the initial production of noise—dividing the flow stream into multiple paths and using several flow resistances in series. Sound pressure level L_p is proportional to mass flow and is dependent on vena contracta velocity. If each path is an independent source, it is easy to show from Eq. (8-120) that p_S^2 is inversely proportional to the number of passages; additionally, smaller passage size shifts the predominate spectral content to higher frequencies, where structural resonance may be less of a problem. Series resistances or multiple stages can reduce maximum velocity and/or produce backpressure to keep jets issuing from multiple passages from acting independently. While some of the basic principles are understood, predicting noise for a particle flow passage requires some empirical data as a basis. Valve manufacturers have developed noise prediction methods for the valves they build. ANSI/ISA-75.17 is a public-domain methodology for standard (non-low-noise) valve types, although treatment of some multistage, multipath types is underway. Low-noise hardware consists of special cages in linear stem valves, perforated domes or plates and multichannel inserts in rotary valves, and separate devices that use multiple fixed restrictions.

Cavitation and Flashing From the discussion of pressure recovery it was seen that the pressure at the vena contracta can be much lower than the downstream pressure. If the pressure on a liquid falls below its vapor pressure p_v, the liquid will vaporize. Due to the effect of surface tension, this vapor phase will first appear as bubbles. These bubbles are carried downstream with the flow, where they collapse if the pressure recovers to a value above p_v. This pressure-driven process of vapor bubble formation and collapse is known as cavitation.

Cavitation has three negative side effects in valves—noise and vibration, material removal, and reduced flow. The bubble collapse process is a violent asymmetric implosion that forms a high-speed microjet and induces pressure waves in the fluid. This hydrodynamic noise and the mechanical vibration that it can produce are far stronger than other noise generation sources in liquid flows. If implosions occur adjacent to a solid component, minute pieces of material can be removed, which, over time, will leave a rough, cinderlike surface.

The presence of vapor in the vena contracta region puts an upper limit on the amount of liquid that will pass through a valve. A mixture of vapor and liquid has a lower density than that of the liquid alone. While Eq. (8-111) is not applicable to two-phase flows because pressure changes are redistributed due to varying density and the two phases do not necessarily have the same average velocity, it does suggest that lower density reduces the total mass flow rate. Figure 8-77 illustrates a typical flow rate/pressure drop relationship. As with compressible gas flow at a given p_1, flow increases as p_2 is decreased until the flow chokes (i.e., no additional fluid will pass). The transition between incompressible and choked flow is gradual because, within the convoluted flow passages of valves, the pressure is actually an uneven distribution at each cross section and consequently vapor formation zones increase gradually. In fact, isolated zones of bubble formation or incipient cavitation often occur at pressure drops well below that at which a reduction in flow is noticeable. The similarity between liquid and gas choking is not serendipitous; it is surmised that the two-phase fluid is traveling at the mixture's sonic velocity in the throat when choked. Complex fluids with components having varying vapor pressures and/or entrained noncondensable gases (e.g., crude oil) will exhibit soft vaporization/implosion transitions.

There are several methods to reduce cavitation or at least its negative side effects. Material damage is slowed by using harder materials and by directing the cavitating stream away from passage walls (e.g., with an angle body flowing down). Sometimes the system can be designed to place the valve in a higher p_2 location or add downstream resistance, which creates backpressure. A low recovery valve has a higher minimum pressure for a given p_2 and so is a means to eliminate the cavitation itself, not just its side effects. In Fig. 8-75, if $p_v < B$, neither valve will cavitate substantially. For $p_v > B$ but $p_v < A$, the high

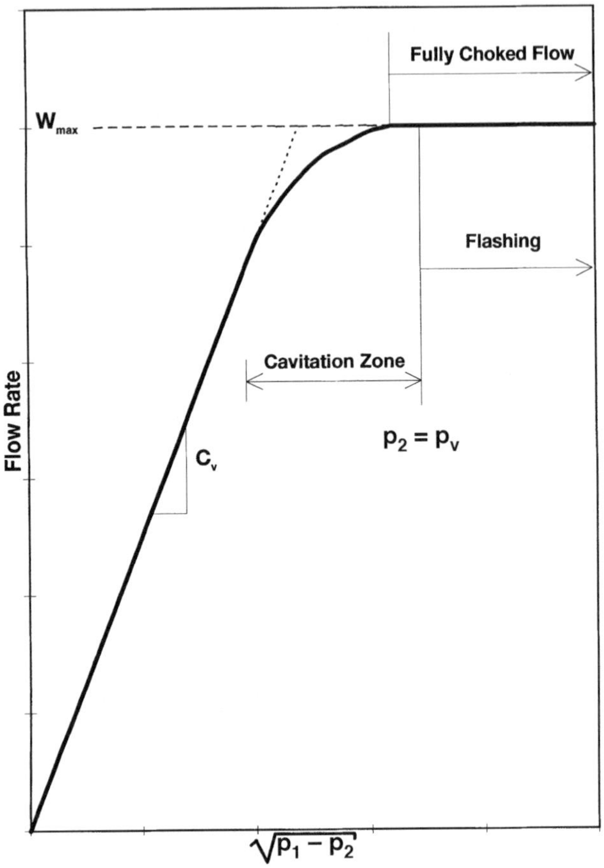

FIG. 8-77 Liquid flow rate versus pressure drop (assuming constant p_1 and p_v).

recovery valve will cavitate substantially, but the low recovery valve will not. Special anticavitation trims are available for globe and angle valves and more recently for some rotary valves. These trims use multiple contraction/expansion stages or other distributed resistances to boost F_L to values sometimes near unity.

If p_2 is below p_v, the two-phase mixture will continue to vaporize in the body outlet and/or downstream pipe until all liquid phase is gone, a condition known as flashing. The resulting huge increase in specific volume leads to high velocities, and any remaining liquid droplets acquire much of the higher vapor-phase velocity. Impingement of these droplets can produce material damage, but it differs from cavitation damage because it exhibits a smooth surface. Hard materials and directing the two-phase jets away from solid surfaces are means to avoid this damage.

Seals, Bearings, and Packing Systems In addition to their control function, valves often need to provide shutoff. FCI 70-2-1998 and IEC 60534-4 recognize six standard classifications and define their as-shipped qualification tests. Class I is an amount agreed to by user and supplier with no test needed. Classes II, III, and IV are based on an air test with maximum leakage of 0.5 percent, 0.1 percent, and 0.01 percent of rated capacity, respectively. Class V restricts leakage to 5×10^{-6} mL of water per second per millimeter of port diameter per bar differential. Class VI allows 0.15 to 6.75 mL/min of air to escape depending on port size; this class implies the need for interference-fit elastomeric seals. With the exception of class V, all classes are based on standardized pressure conditions that may not represent actual conditions. Therefore, it is difficult to estimate leakage in service. Leakage normally increases over time as seals and seating surfaces become nicked or worn. Leak passages across the seat-contact line, known as wire drawing, may form and become worse over time—even in hard metal seats under sufficiently high-pressure differentials.

Polymers used for seat and plug seals and internal static seals include PTFE (polytetrafluoroethylene) and other fluorocarbons, polyethylene, nylon, polyether-ether-ketone, and acetal. Fluorocarbons are often carbon- or glass-filled to improve mechanical properties and heat resistance. Temperature and chemical compatibility with the process fluid are the key selection criteria. Polymer-lined bearings and guides are used to decrease friction, which lessens dead band and reduces actuator force requirements. See Sec. 25, "Materials of Construction," for properties.

Packing forms the pressure-tight seal, where the stem protrudes through the pressure boundary. Packing is typically made from PTFE or, for high temperature, a bonded graphite. If the process fluid is toxic, more sophisticated systems such as dual packing, live-loaded, or a flexible metal bellows may be warranted. Packing friction can significantly degrade control performance. Pipe, bonnet, and internal-trim joint gaskets are typically a flat sheet composite. Gaskets intended to absorb dimensional mismatch are typically made from filled spiral-wound flat stainless-steel wire with PTFE or graphite filler. The use of asbestos in packing and gaskets has been largely eliminated.

Flow Characteristics The relationship between valve flow and valve travel is called the valve flow characteristic. The purpose of flow characterization is to make loop dynamics independent of load, so that a single controller tuning remains optimal for all loads. Valve gain is one factor affecting loop dynamics. In general, gain is the ratio of change in output to change in input. The input of a valve is travel y, and the output is flow w. Since pressure conditions at the valve can depend on flow (hence travel), valve gain is

$$\frac{dw}{dy} = \frac{\partial w}{\partial C_V}\frac{dC_V}{dy} + \frac{\partial w}{\partial p_1}\frac{dp_1}{dy} + \frac{\partial w}{\partial p_2}\frac{dp_2}{dy} \qquad (8\text{-}121)$$

An inherent valve flow characteristic is defined as the relationship between flow rate and travel, under constant-pressure conditions. Since the rightmost two terms in Eq. (8-121) are zero in this case, the inherent characteristic is necessarily also the relationship between flow coefficient and travel.

Figure 8-78 shows three common inherent characteristics. A linear characteristic has a constant slope, meaning the inherent valve gain is a constant. The most popular characteristic is equal-percentage, which gets its name from the fact that equal changes in travel produce equal-percentage changes in the existing flow coefficient. In other words, the slope of the curve is proportional to C_V, or equivalently that inherent valve gain is proportional to flow. The equal-percentage characteristic can be expressed mathematically by

$$C_V(y) = (\text{rated } C_V)\exp\left[\left(\frac{y}{\text{rated } y} - 1\right)\ln R\right] \qquad (8\text{-}122)$$

This expression represents a set of curves parameterized by R. Note that C_V ($y = 0$) equals (rated C_V)/R rather than zero; real equal-percentage characteristics deviate from theory at some small travel to meet shutoff requirements. An equal-percentage characteristic provides perfect compensation for a process where the gain is inversely proportional to flow (e.g., liquid pressure). *Quick opening* does not have a standardized mathematical definition. Its shape arises naturally from high-capacity plug designs used in on/off service globe valves. Frequently, pressure conditions at the valve will change with flow rate. This so-called process influence [the rightmost two terms on the right-hand side of Eq. (8-121)] combines with inherent gain to express the installed valve gain. The flow versus travel relationship for a specific set of conditions is called the installed flow characteristic. Typically, valve Δp decreases with load, since pressure losses in the piping system increase with flow. Figure 8-79 illustrates how allocation of total system head to the valve influences the installed flow characteristics. For a linear or quick-opening characteristic, this transition toward a concave down shape would be more extreme. This effect of typical process pressure variation, which causes equal-percentage

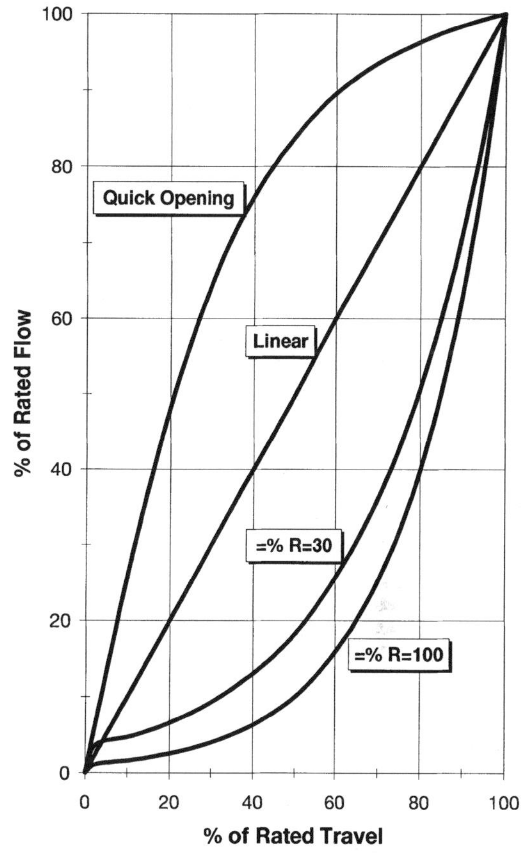

FIG. 8-78 Typical inherent flow characteristics.

characteristics to have fairly constant installed gain, is one reason the equal-percentage characteristic is the most popular.

Due to clearance flow, flow force gradients, seal friction, and the like, flow cannot be throttled to an arbitrarily small value. Installed rangeability is the ratio of maximum to minimum controllable flow. The actuator and positioner, as well as the valve, influence the installed rangeability. Inherent rangeability is defined as the ratio of the largest to the smallest C_V within which the characteristic meets specified criteria (see ISA 75.11). The R value in the equal-percentage definition is a theoretical rangeability only. While high installed rangeability is desirable, it is also important not to oversize a valve; otherwise, turndown (ratio of maximum normal to minimum controllable flow) will be limited.

Sliding stem valves are characterized by altering the contour of the plug when the port and plug determine the minimum (controlling) flow area. Passage area versus travel is also easily manipulated in characterized cage designs. Inherent rangeability varies widely, but typical values are 30 for contoured plugs and 20 to 50 for characterized cages. While these types of valves can be characterized, the degree to which manufacturers conform to the mathematical ideal is revealed by plotting measured C_V versus travel. Note that ideal equal-percentage will plot as a straight line on a semilog graph. Custom characteristics that compensate for a specific process are possible.

Rotary stem-valve designs are normally offered only in their naturally occurring characteristic, since it is difficult to appreciably alter this. If additional characterization is required, the positioner or controller may be characterized. However, these approaches are less direct, since it is possible for device nonlinearity and dynamics to distort the compensation.

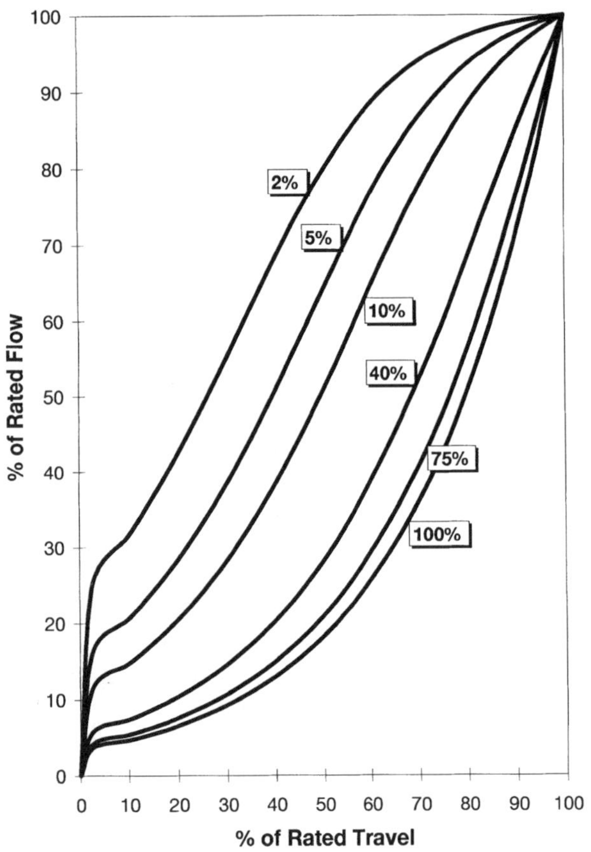

FIG. 8-79 Installed flow characteristic as a function of percent of total system head allocated to the control valve (assuming constant-head pump, no elevation head loss, and an R equal to 30 equal-percentage inherent characteristic).

FIG. 8-80 Valve and actuator with valve positioner attached. (*Courtesy Fisher Controls International LLC.*)

VALVE CONTROL DEVICES

Devices mounted on the control valve that interface various forms of input signals, monitor and transmit valve position, or modify valve response are valve control devices. In some applications, several auxiliary devices are used together on the same control valve. For example, mounted on the control valve, one may find a current-to-pressure transducer, a valve positioner, a volume booster relay, a solenoid valve, a trip valve, a limit switch, a process controller, and/or a stem position transmitter. Figure 8-80 shows a valve positioner mounted on the yoke leg of a spring and diaphragm actuator.

As most throttling control valves are still operated by pneumatic actuators, the control valve device descriptions that follow relate primarily to devices that are used with pneumatic actuators. The functions of hydraulic and electrical counterparts are very similar. Specific details on a particular valve control device are available from the vendor of the device.

Valve Positioners The valve positioner, when combined with an appropriate actuator, forms a complete closed-loop valve position control system. This system makes the valve stem conform to the input signal coming from the process controller in spite of force loads that the actuator may encounter while moving the control valve. Usually, the valve positioner is contained in its own enclosure and is mounted on the control valve.

The key parts of the positioner/actuator system, shown in Fig. 8-81*a*, are (1) an input conversion network, (2) a stem position feedback network, (3) a summing junction, (4) an amplifier network, and (5) an actuator.

The input conversion network shown is the interface between the input signal and the summer. This block converts the input current or pressure (from an I/P transducer or a pneumatic process controller) to a voltage, electric current, force, torque, displacement, or other particular variable that can be directly used by the summer. The input conversion usually contains a means to adjust the slope and offset of the block to provide for a means of spanning and zeroing the positioner during calibration. In addition, means for changing the sense (known as "action") of the input/output characteristic are often addressed in this block. Also exponential, logarithmic, or other predetermined characterization can be put in this block to provide a characteristic that is useful in offsetting or reinforcing a nonlinear valve or process characteristic.

The stem position feedback network converts stem travel to a useful form for the summer. This block includes the feedback linkage which varies with actuator type. Depending on positioner design, the stem position feedback network can provide span and zero and characterization functions similar to that described for the input conversion block.

The amplifier network provides signal conversion and suitable static and dynamic compensation for good positioner performance. Control from this block usually reduces to a form of proportional or proportional plus derivative control. The output from this block in the case of a pneumatic positioner is a single connection to the spring and diaphragm actuator or two connections for push/pull operation of a

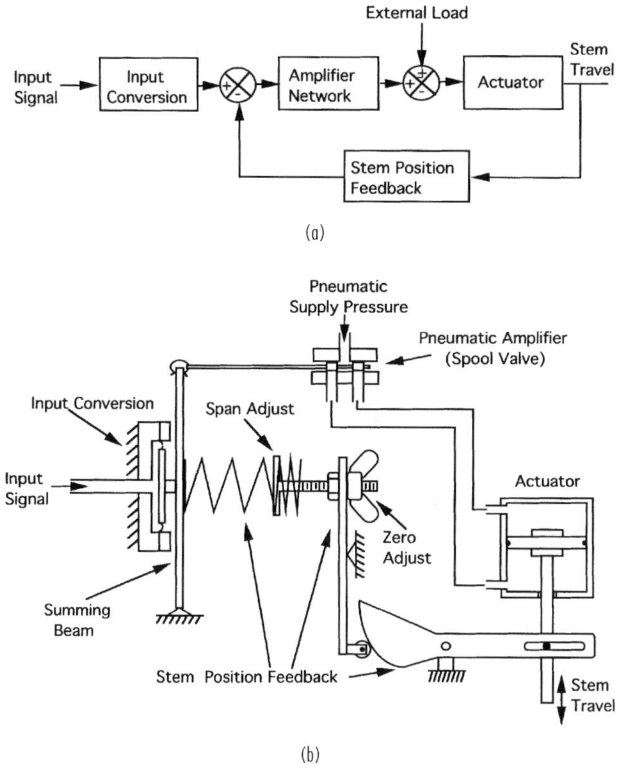

FIG. 8-81 Positioner/actuators. (*a*) Generic block diagram. (*b*) Example of a pneumatic positioner/actuator.

springless piston actuator. The action of the amplifier network and the action of the stem position feedback can be reversed together to provide for reversed positioner action.

By design, the gain of the amplifier network shown in Fig. 8-81*a* is made very large. Large gain in the amplifier network means that only a small proportional deviation will be required to position the actuator through its active range of travel. This means that the signals into the summer track very closely and that the gain of the input conversion block and the stem position feedback block determine the closed-loop relationship between the input signal and the stem travel.

Large amplifier gain also means that only a small amount of additional stem travel deviation will result when large external force loads are applied to the actuator stem. For example, if the positioner's amplifier network has a gain of 50 (and assuming that high packing box friction loads require 25 percent of the actuator's range of thrust to move the actuator), then only 25 percent/50 (or 0.5 percent deviation) between input signal and output travel will result due to valve friction.

Figure 8-81*b* is an example of a pneumatic positioner/actuator. The input signal is a pneumatic pressure that (1) moves the summing beam, which (2) operates the spool valve amplifier, which (3) provides flow to and from the piston actuator, which (4) causes the actuator to move and continue moving until (5) the feedback force returns the beam to its original position and stops valve travel at a new position. Typical positioner operation is thereby achieved.

Static performance measurements related to positioner/actuator operation include the conformity, measured accuracy, hysteresis, dead band, repeatability, and locked stem pressure gain. Definitions and standardized test procedures for determining these measurements can be found in ISA-S75.13, "Method of Evaluating the Performance of Positioners with Analog Input Signals and Pneumatic Output."

Dynamics of Positioner-Based Control Valve Assemblies Control valve assemblies are complete, functional units that include the valve body, actuator, positioner, if so equipped, associated linkages, and any auxiliary equipment such as current to pneumatic signal transducers and air supply pressure regulators. Although performance information such as frequency response, sensitivity, and repeatability data may be available for a number of these components individually, it is the performance of the entire assembly that will ultimately determine how well the demand signal from the controller output is transferred through the control valve to the process. The valve body, actuator, and positioner combination is typically responsible for the majority of the control valve assembly's dynamic behavior. On larger actuators, the air supply pressure regulator capacity or other airflow restrictions may limit the control valve assembly's speed of response.

The control valve assembly response can usually be characterized quite well by using a first-order plus dead-time response model. The control valve assembly will also exhibit backlash, stiction, and other nonlinear behavior. During normal operation of a control loop, the controller usually makes small output changes from one second to the next. Typically this change is less than 1 percent. With very small controller output changes, e.g., less than 0.1 percent, the control valve assembly may not move at all. As the magnitude of the controller output change increases, eventually the control valve will move. At the threshold of movement, the positional accuracy and repeatability of the control valve are usually quite poor. The speed of response may be quite slow and may occur after a number of seconds of dead time. This poor performance is due to the large backlash and stiction effects relative to the requested movement and the small output change of the positioner. With a further increase in the magnitude of the controller output steps, the behavior of the control valve typically becomes more repeatable and "linear." Dead time usually drops to only a fraction of a second, and the first-order time constant becomes faster. For much larger steps in the controller output, e.g., over 10 percent, the positioner and air supply equipment may be unable to deliver the necessary air volume to maintain the first-order response. In this case, the control valve will exhibit very little dead time, but will be rate-limited and will ramp toward the requested position. It is within the linear region of motion that the potential for the best control performance exists.

When one is specifying a control valve for process control applications, in addition to material, style, and size information, the dynamic response characteristics and maximum allowable dead band (sum of backlash, stiction, and hysteresis effects) *must* be stated. The requirement for the control valve assembly's speed of response is ultimately determined by the dynamic characteristics of the process and the control objectives. Typically, the equivalent first-order time constant specified for the control valve assembly should be at least 5 times faster than the desired controller closed-loop time constant. If this requirement is not met, the tuning of the control loop must be slowed down to accommodate the slow control valve response, otherwise, control robustness and stability may be compromised. The dead band of the control valve assembly is typically the determining factor for control resolution and frequently causes control instability in the form of a "limit" cycle. The controller output will typically oscillate across a range that is 1 to 2 times the magnitude of the control valve dead band. This is very dependent on the nature of the control valve nonlinearities, the process dynamics, and the controller tuning. The magnitude of the process limit cycle is determined by the size of the control valve dead band multiplied by the installed gain of the control valve. For this reason, a high-performance control valve assembly, e.g., with only 0.5 percent dead band, may cause an unacceptably large process limit cycle if the valve is oversized and has a high installed gain. For typical process control applications, the installed gain of the control valve should be in the range of 0.5 to 2 percent of the process variable span per percent of the controller output. The total dead band of the control valve assembly should be less than 1 percent. For applications that require more precise control, the dead band and possibly the installed gain of the control valve must be reduced. Specialized actuators are available that are accurate down to 0.1 percent or less. At this level of performance, however, the design of the valve body, bearings, linkages, and seals starts to become a significant source of dead band.

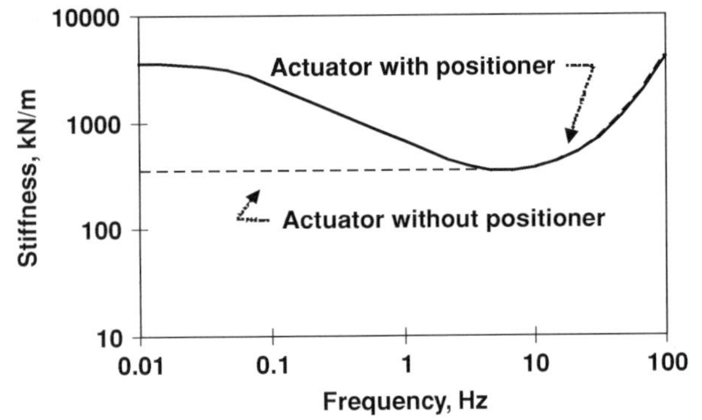

FIG. 8-82 Actuator stiffness as a function of frequency for a 69-in^2 spring and diaphragm pneumatic actuator. Actuator with positioner exhibits higher stiffness over the lower frequency range compared to that of the pneumatic actuator without a positioner.

Positioner/Actuator Stiffness Minimizing the effect of dynamic loads on valve stem travel is an important characteristic of the positioner/actuator. Stem position must be maintained in spite of changing reaction forces caused by valve throttling. These forces can be random (buffeting force) or can result from a negative-slope force/stem travel characteristic (negative gradient); either could result in valve stem instability and loss of control. To reduce and eliminate the effect of these forces, the effective stiffness of the positioner/actuator must be made sufficiently high to maintain stationary control of the valve stem.

The stiffness characteristic of the positioner/actuator varies with the forcing frequency. Figure 8-82 indicates the stiffness of the positioner/actuator is increased at low frequencies and is directly related to the locked-stem pressure gain provided by the positioner. As frequency increases, a dip in the stiffness curve results from dynamic gain attenuation in the pneumatic amplifiers in the positioner. The value at the bottom of the dip is the sum of the mechanical stiffness of the spring in the actuator and the air spring effect produced by air enclosed in the actuator casing. At yet higher frequencies, actuator inertia dominates and causes a corresponding rise in system stiffness.

The air spring effect results from adiabatic expansion and compression of air in the actuator casing. Numerically, the small perturbation value for air spring stiffness in newtons per meter is given by

$$\text{Air spring rate} = \frac{\gamma p_a A_a^2}{V} \tag{8-123}$$

where γ is the ratio of specific heats (1.4 for air), p_a is the actuator pressure in pascals absolute, A_a is the actuator pressure area in square meters, and V is the internal actuator volume in cubic meters.

Positioner Application Positioners are widely used on pneumatic valve actuators. Often they provide improved process loop control because they reduce valve-related nonlinearity. Dynamically, positioners maintain their ability to improve control valve performance for sinusoidal input frequencies up to about one-half of the positioner bandwidth. At input frequencies greater than this, the attenuation in the positioner amplifier network gets large, and valve nonlinearity begins to affect final control element performance more significantly. Because of this, the most successful use of the positioner occurs when the positioner response bandwidth is greater than twice that of the most dominant time lag in the process loop.

Some typical examples in which the dynamics of the positioner are sufficiently fast to improve process control are the following:

1. *In a distributed control system (DCS) process loop with an electronic transmitter.* The DCS controller and the electronic transmitter have time constants that are dominant over the positioner response. Positioner operation is therefore beneficial in reducing valve-related nonlinearity.

2. *In a process loop with a pneumatic controller and a large process time constant.* Here the process time constant is dominant, and the positioner will improve the linearity of the final control element. Some common processes with large time constants that benefit from positioner application are liquid level, temperature, large-volume gas pressure, and mixing.

3. *Additional situations in which valve positioners are used:*

a. On springless actuators where the actuator is not usable for throttling control without position feedback.

b. When split ranging is required to control two or more valves sequentially. In the case of two valves, the smaller control valve is calibrated to open in the lower half of the input signal range, and a larger valve is calibrated to open in the upper half of the input signal range. Calibrating the input command signal range in this way is known as split-range operation and increases the practical range of throttling process flows over that of a single valve.

c. In open-loop control applications where best static accuracy is needed. On occasion, positioner use can degrade process control. Such is the case when the process controller, process, and process transmitter have time constants that are similar to or smaller than that of the positioner/actuator. This situation is characterized by low process controller proportional gain (gain < 0.5), and hunting or limit cycling of the process variable is observed.

Improvements here can be made by doing one of the following:

1. Install a dominant first-order, low-pass filter in the loop ahead of the positioner and retune the process loop. This should allow increased proportional gain in the process loop and reduce hunting. Possible means for adding the filter include adding it to the firmware of the DCS controller, by adding an external *RC* network on the output of the process controller or by enabling the filter function in the input of the positioner, if it is available. Also, some transducers, when connected directly to the actuator, form a dominant first-order lag that can be used to stabilize the process loop.

2. Select a positioner with a faster response characteristic.

Processor-Based Positioners When designed around an electronic microcontroller, the valve positioner [now commonly referred to as a *digital valve controller* (DVC)] takes on additional functionality that provides convenience and performance enhancements over the traditional design. The most common form of processor-based positioner, shown in Fig. 8-81, is a digitally communicating stem position controller that operates by using the fundamental blocks shown in Fig. 8-81*a*. A local display is part of the positioner and provides tag information, command input and travel, servo tuning parameters, and diagnostic information. Often auxiliary sensors are integrated into the device to provide increased levels of functionality and performance. Sensed variables can include actuator pressure, relay input pressure, relay valve position, board temperature, or a discrete input. A 4- to 20-mA valve travel readback circuit is also common. The travel sensor is based on a potentiometer or can be a noncontacting type such as a variable capacitance sensor, Hall effect sensor, or GMR device. Some positioners require a separate connection to an ac or dc supply voltage, but the majority of the designs are "loop-powered," which means that they receive power either through the current input (for positioners that require a 4- to 20-mA analog input signal) or through the digital communications link when the control signal is a digital signal.

Processor-based positioners support automatic travel calibration and automatic response tuning for quick commissioning of the final control element. Features of this type of valve positioner include compensators for improved static and dynamic travel response; diagnostics for evaluating positioner, actuator, and valve health; and the capability to be polled from remote locations through a PC-based application or through a handheld communicator attached to the field wiring. Capability to support custom firmware for special valve applications, such as emergency safety shutdown, is also a characteristic of the processor-based design.

Digital Field Communications To provide increased data transmission capability between valve-mounted devices and the host control system, manufacturers are providing digital network means in their devices. The field networks, commonly known as field buses, compete fiercely in the marketplace and have varying degrees of flexibility and specific application strengths. A prospective field bus customer is advised to study the available bus technologies and to make a selection based on needs, and not be seduced by the technology itself.

Generally, a field bus protocol must be nonproprietary ("open") so that different vendors of valve devices can design their bus interface to operate properly on the selected field bus network. Users demand that the devices be "interoperable" so that the device will work with other devices on the same segment or can be substituted with a device from an alternate manufacturer. International standardization of some of the protocols is currently underway (for example, IEC 61158) whereas others are sponsored by user groups or foundations that provide democratic upgrades to the standard and provide network compliance testing.

The physical wiring typically used is the plant standard twisted-pair wiring for 4- to 20-mA instrumentation. Because of the networking capability of the bus, more than one device can be supported on a single pair of wires, and thus wiring requirements are reduced. Compared to a host level bus such as Ethernet, field buses exhibit slower communication rates, have longer transmission distance capability (1 to 2 km), use standard two-wire installation, are capable of multidrop busing, can support bus-powered devices, do not have redundant modes of bus operation, and are available for intrinsically safe installations. Devices on the field bus network may be either powered by the bus itself or powered separately.

The simplest digital networks available today support discrete sensors and on/off actuators, including limit switches and motor starters. Networks of this type have fast cycle times and are often used as an alternative to PLC discrete I/O. More sophisticated field networks are designed to support process automation, more complex process transmitters, and throttling valve actuators. These process-level networks are fundamentally continuous and analoglike in operation, and data computation is floating-point. They support communication of materials of construction, calibration and commissioning, device and loop level diagnostics (including information displays outlining corrective action), and unique manufacturer-specific functionality. Some process networks are able to automatically detect, identify, and assign an address to a new device added to the network, thus reducing labor, eliminating addressing errors, and indicating proper network function immediately after the connection is made. Final control elements operated by the process-level network include I/P transducers, motorized valves, digital valve controllers, and transmitters.

A particular field network protocol known as HART®* (Highway Addressable Remote Transducer) is the most widely used field network protocol. It is estimated that as of 2004 there are more than 14 million HART-enabled devices installed globally and that 70 percent of all processor-based process measurement and control instruments installed each year use HART communications. HART's popularity is based on its similarity to the traditional 4- to 20-mA field signaling and thus represents a safe, controlled transition to digital field communications without the risk often associated with an abrupt change to a totally digital field bus. With this protocol, the digital communications occur over the same two wires that provide the 4- to 20-mA process control signal without disrupting the process signal. The protocol uses the frequency-shift keying (FSK) technique (see Fig. 8-83) where two individual frequencies, one representing the mark and the other representing the space, are superimposed on the 4- to 20-mA current signal. As the average value of the signals used is zero, there is no dc offset value added to the 4- to 20-mA signal. The HART protocol is principally a master/slave protocol which means that a field device (slave) speaks only when requested by a master device. In this mode of operation, the slave can update the master at a rate of twice per second. An optional communication mode, *burst mode,* allows a HART slave device to continuously broadcast updates without stimulus requests from the master device. Update rates of 3 to 4 updates per second are typical in the burst mode of operation.

HART-enabled devices are provided by the valve device manufacturer at little or no additional cost. The HART network is compatible with existing 4- to 20-mA applications using current plant personnel and practices, provides for a gradual transition from analog to fully digital protocols, and is provided by the valve device manufacturer at little or no additional cost. Contact the HART Communication Foundation for additional information.

Wireless digital communication to and from the final control element is not yet commercially available but is presently being investigated by more than one device manufacturer. The positive attribute of a wireless field network is the reduced cost of a wireless installation compared to a wired installation. Hurdles for wireless transmissions include security from nonnetwork sources, transmission reliability in the plant environment, limited bus speed, and the conservative nature of the process industry relative to change. Initial installations of wireless networks will support secondary variables and diagnostics, then primary control of processes with large time constants, and finally general application to process control. Both point-to-point and mesh architectures are being evaluated for commercialization at the device level. Mesh architectures rely on the other transmitting devices in the area to receive and then pass on any data transmission, thus rerouting communications around sources of interference. Two unlicensed spread spectrum radio bands are the main focus for current wireless development: 900 MHz and 2.4 GHz. The 900-MHz band is unique to North America and has better propagation and penetrating properties than the 2.4-GHz band. The 2.4-GHz band is a worldwide band and has wider channels, allowing much higher data rates. The spread

*HART is a registered trademark of the HART Communication Foundation.

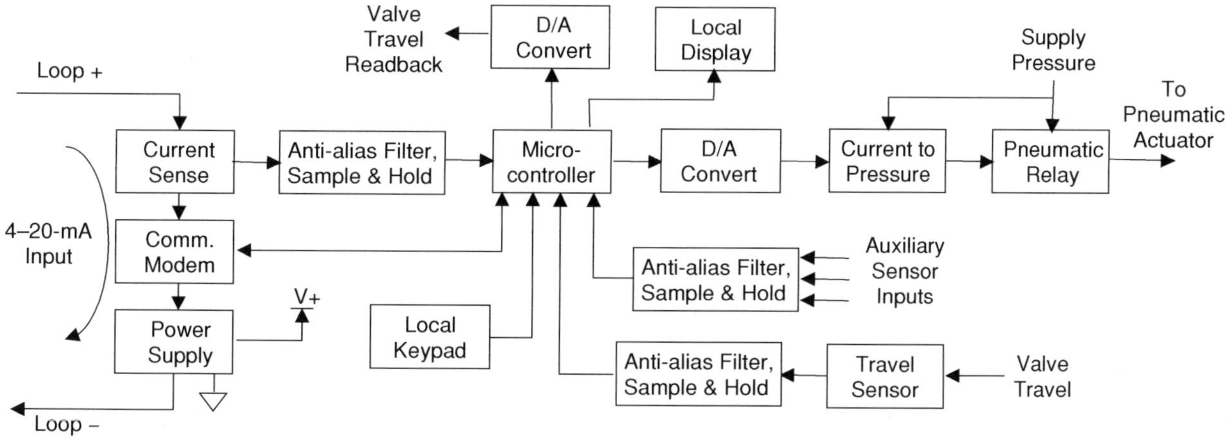

FIG. 8-83 Generic loop powered digital valve controller.

spectrum technique uses multiple frequencies within the radio band to transmit data. Spread spectrum is further divided into the direct sequence technique, where the device changes frequency many times per data bit, and the frequency-hopping technique, where the device transmits a data packet on one frequency and then changes to a different frequency. Because of the rapid growth expected in this decade, the prospective wireless customer is encouraged to review up-to-date literature to determine the state of field wireless commercialization as it applies to her or his specific application.

Diagnostic Capability The rapid proliferation of communicating, processor-based digital valve controllers over the last decade has led to a corresponding rise in diagnostic capability at the control valve. Diagnosing control valve health is critical to plant operation as maintenance costs can be reduced by identifying the valves that are candidates for repair. Less time is spent during plant shutdown repairing valves that do not need repair, which ultimately results in increased online operating time. Valve diagnostics can detect and flag a failed valve more quickly than by any other means, and can be configured to cause the valve to move to its fail-safe position on detection of specified fault conditions. The diagnostic-enabled positioner, when used with its host-based software application, can pinpoint exact components in a given final control element that have failed, and can recommend precise maintenance procedures to follow to remedy the fault condition.

The state variables that provide valve position control are used to diagnose the health of the final control element. In addition, some digital valve controller designs integrate additional sensors into their construction to provide increased diagnostic capability. For example, pressure sensors are provided to detect supply pressure, actuator pressure (upper and lower cylinder pressures in the case of a springless piston actuator), and internal pilot pressure. Also, the position of the pneumatic relay valve is available in some designs to provide quiescent flow data used for leak detection in the actuator.

Valve diagnostics are divided into two types: online and offline. Offline diagnostics are those diagnostics that occur when the control valve is bypassed or otherwise isolated from the process. The offline diagnostic routine manipulates the travel command to the valve and records the corresponding valve travel, actuator pressure, and servodrive value. These parameters are plotted in various combinations to provide hysteresis plus dead-band information, actuator operating pressure as a function of travel, valve friction, servodrive performance, valve seating load, positioner calibration endpoints, and dynamic response traces. Small- and large-amplitude step inputs as well as large slow ramps (exceeding 100 percent of the input range) are common offline test waveforms generated by the diagnostic as command inputs for offline diagnostic tests. Figure 8-84 is an example of one offline diagnostic test performed on a small globe valve actuated by a spring and diaphragm actuator. During this test the command input, travel, actuator pressure, and servodrive level are recorded and plotted as they result from a command input that is slowly ramped by the diagnostic routine (Fig. 8-85*a*). This diagnostic is extremely useful in detecting problems with the valve/actuator system and can flag potential problems with the final control element before catastrophic failure occurs. For example, Fig. 8-85*b* indicates the overall tracking capability of the control valve, and Fig. 8-85*c* indicates the pressure operating range of the actuator and the amount of frictional force resulting from the combined effects of valve packing and valve plug contact. Figure 8-85*d* displays the level of servodrive required to stroke the valve from one end of travel to the other. The composite operative health of the control valve is determined through comparison of the empirical levels presented in Fig. 8-85 with the manufacturers' recommendations. Recommended maintenance actions result from this comparison.

Online diagnostics are diagnostics that monitor and evaluate conditions at the control valve during normal throttling periods (i.e., during valve-in-service periods). Online diagnostics monitor mean levels and disturbances generated in the normal operation of the valve and typically do not force or generate disturbances on the valve's operation. For example, an online diagnostic can calculate travel deviation relative to the input command and flag a condition where the valve travel has deviated beyond a preset band. Such an event, if it exists for more than a short time, indicates that the valve has lost its ability to track the input

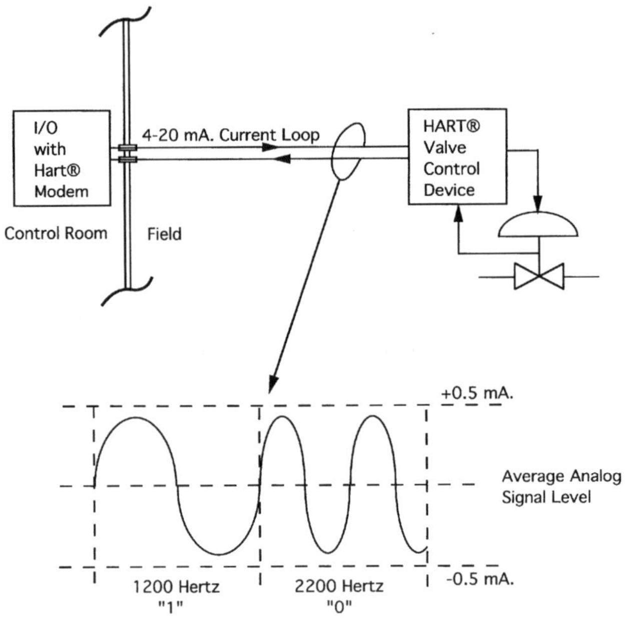

FIG. 8-84 Hybrid point-to-point communications between the control room and the control valve device.

command within specified limits. Additional diagnostics could suggest that the feedback linkage has ceased functioning, or that the valve has stuck, or that some other specific malfunction is the cause of excess travel deviation. The manufacturer of the positioner diagnostic incorporates default limits into the host software application that are used to determine the relative importance of a specific deviation. To quickly indicate the severity of a problem detected by a diagnostic routine, a red, yellow, or green, or "advise, maintenance now, or failed," indication is presented on the user-interface screen for the valve problem diagnosed. Help notes and recommended remedial action are available by pointing and clicking on the diagnostic icon presented on the user's display.

Event-triggered recording is an online diagnostic technique supported in digital valve controllers (DVCs). Functionally a triggering event, such as a valve coming off a travel stop or a travel deviation alert, starts a time-series recording of selected variables. A collection of variables such as the input command, stem travel, actuator pressure, and drive command are stored for several minutes before and after the triggered event. These variables are then plotted as time series for immediate inspection or are stored in memory for later review. Event-triggering diagnostics are particularly useful in diagnosing valves that are closed or full-open for extended periods. In this case the event-triggered diagnostic focuses on diagnostic rich data at the time the valve is actually in operation and minimizes the recording of flat-line data with little diagnostic content. Other online diagnostics detected by DVC manufacturers include excess valve friction, supply pressure failure, relay operation failure, broken actuator spring, current to pressure module failure, actuator diaphragm leaking, and shifted travel calibration.

Safety shutdown valves, which are normally wide open and operate infrequently, are expected to respond to a safety trip command reliably and without fault. To achieve the level of reliability required in this application, the safety valve must be periodically tested to ensure positive operation under safety trip conditions. To test the operation of the shutdown system without disturbing the process, the traditional method is to physically lock the valve stem in the wide-open position and then to electrically operate the pneumatic shutdown solenoid valve. Observing that the pneumatic solenoid valve has properly vented the actuator pressure to zero, the actuator is seen as capable of applying sufficient spring force to close the valve, and a positive safety valve test is indicated. The

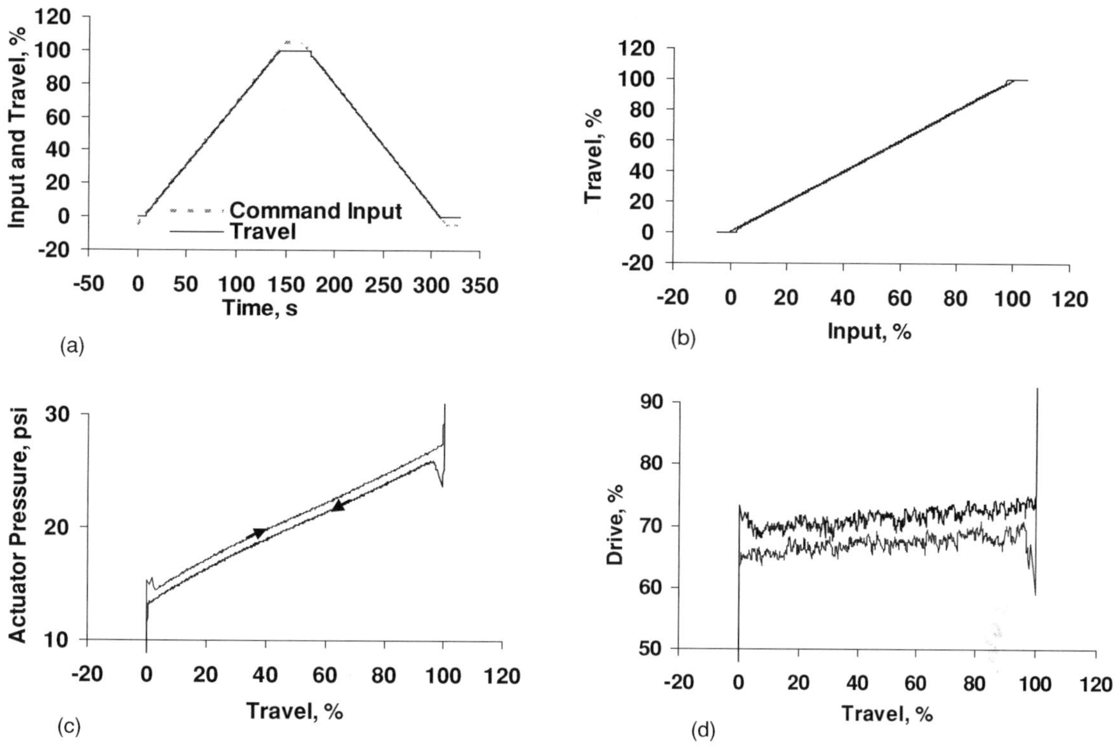

FIG. 8-85 Offline valve diagnostic scan showing results of a diagnostic ramp. (*a*) The command input and resulting travel. (*b*) The dynamic scan. (*c*) The valve signature. (*d*) The servodrive versus travel plot. The hysteresis shown in the valve signature results from sliding friction due to valve packing and valve plug contact.

pneumatic solenoid valve is then returned to its normal electrical state, the actuator pressure returns to full supply pressure, and the valve stem lock mechanism is removed. This procedure, though necessary to enhance process safety, is time-consuming and takes the valve out of service during the locked stem test. Digital valve controllers are able to validate the operation of a safety shutdown valve by using an online diagnostic referred to as a partial stroke test. The partial stroke test is substituted for the traditional test method described above and does not require the valve to be locked in the wide-open position to perform the test. In a fashion similar to that shown in Fig. 8-85*a* (the partial stroke diagnostic), the system physically ramps the command input to the positioner from the wide-open position to a new position, pauses at the new position for a few seconds, and then ramps the command input back to the wide-open position (see Fig. 8-86*a*). During this time, the valve travel measurement is monitored and compared to the input command. If the travel measurement deviates for the input by more than a fixed amount for the configured period of time, the valve is considered to have failed the test and a failed-test message is communicated to the host system. Also during this test, the actuator pressure required to move the valve is detected via a dedicated pressure sensor (see Fig. 8-86*b*). If the thrust (pressure) required to move the valve during the partial stroke test exceeds the predefined thrust limit for this test, the control valve is determined to have a serious sticking problem, the test is immediately aborted, and the valve is flagged as needing maintenance. The partial stroke test can be automated to perform on a periodic basis, for instance, once a week; or it can be initialized by operator request at any time. The amount of valve travel that occurs during the partial stroke test is typically limited to a minimum valve position of 70 percent open or greater. This limit is imposed to prevent the partial stroking of the safety valve from significantly affecting the process flow through the valve. Comparison of partial stroke curves from past tests can indicate the gradual degradation of valve components. Use of "overlay" graphics, identification of unhealthy shifts in servodrive, increases in valve friction, and changes in dynamic response provide information leading to a diagnosis of needed maintenance.

In addition to device-level diagnostics, networked final control elements, process controllers, and transmitters can provide "loop" level diagnostics that can detect loops that are operating below expectations. Process variability, time in a limit (saturated) condition, and time in the wrong control mode are metrics used to detect problems in process loop operation.

Transducers The current-to-pressure transducer (I/P transducer) is a conversion interface that accepts a standard 4- to 20-mA input current from the process controller and converts it to a pneumatic output in a standard pneumatic pressure range [normally 0.2 to 1.0 bar (3 to 15 psig) or, less frequently, 0.4 to 2.0 bar (6 to 30 psig)]. The output pressure generated by the transducer is connected directly to the pressure connection on a spring-opposed diaphragm actuator or to the input of a pneumatic valve positioner.

Figure 8-87*a* is the schematic of a basic I/P transducer. The transducer shown is characterized by (1) an input conversion that generates an angular displacement of the beam proportional to the input current, (2) a pneumatic amplifier stage that converts the resulting angular displacement to pneumatic pressure, and (3) a pressure area that serves as a means to return the beam to very near its original position when the new output pressure is achieved. The result is a device that generates a pressure output that tracks the input current signal. The transducer shown in Fig. 8-88*a* is used to provide pressure to small load volumes (normally 4.0 in^3 or less), such as a positioner or booster input. With only one stage of pneumatic amplification, the flow

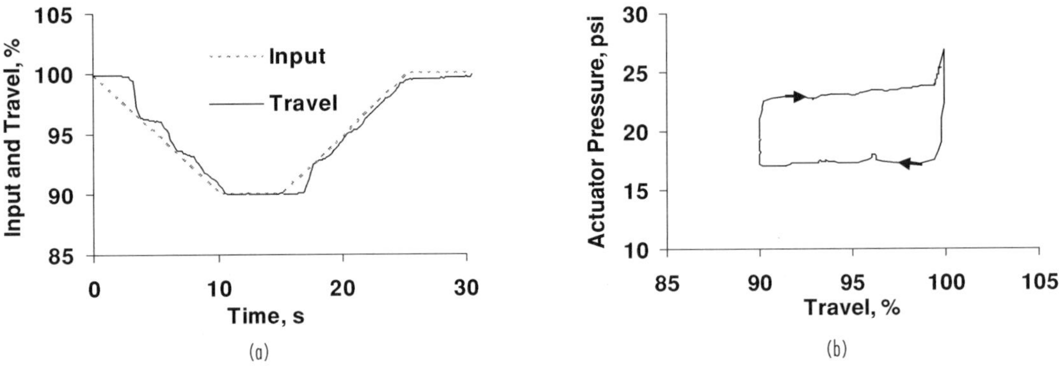

FIG. 8-86 Online partial stroke diagnostic used to validate the operability of a pneumatically operated safety shutdown valve. (*a*) Input command generated by the diagnostic and resulting travel. (*b*) Actuator pressure measured over the tested range of travel.

capacity of this transducer is limited and not sufficient to provide responsive load pressure directly to a pneumatic actuator.

The flow capacity of the transducer can be increased by adding a booster relay such as the one shown in Fig. 8-87*b*. The flow capacity of the booster relay is nominally 50 to 100 times that of the nozzle amplifier shown in Fig. 8-87*a* and makes the combined transducer/booster suitably responsive to operate pneumatic actuators. This type of transducer is stable for all sizes of load volume and produces measured accuracy (see ANSI/ISA-51.1, "Process Instrumentation Terminology," for the definition of measured accuracy) of 0.5 to 1.0 percent of span.

Better measured accuracy results from the transducer design shown in Fig. 8-87*c*. In this design, pressure feedback is taken at the output of the booster relay stage and fed back to the main summer. This allows the transducer to correct for errors generated in the pneumatic booster as well as errors in the I/P conversion stage. Also, particularly with the new analog electric and digital versions of this design, PID control is used in the transducer control network to give extremely good static accuracy, fast dynamic response, and reasonable stability into a wide range of load volumes (small instrument bellows to large actuators). Also environmental factors such as temperature change, vibration, and supply pressure fluctuation affect this type of transducer the least. Even a perfectly accurate I/P transducer cannot compensate for stem position errors generated by friction, backlash, and varying force loads coming from the actuator and valve. To do this compensation, a different control valve device—the valve positioner—is required.

Booster Relays The booster relay is a single-stage power amplifier having a fixed gain relationship between the input and output pressures. The device is packaged as a complete stand-alone unit with pipe thread connections for input, output, and supply pressure. The booster amplifier shown in Fig. 8-87*b* shows the basic construction of the booster relay. Enhanced versions are available that provide specific features such as (1) variable gain to split the output range of a pneumatic controller to operate more than one valve or to provide additional actuator force; (2) low hysteresis for relaying measurement and control signals; (3) high flow capacity for increased actuator stroking speed; and (4) arithmetic, logic, or other compensation functions for control system design.

A particular type of booster relay, called a dead-band booster, is shown in Fig. 8-88. This booster is designed to be used exclusively between the output of a valve positioner and the input to a pneumatic actuator. It is designed to provide extra flow capacity to stroke the actuator faster than with the positioner alone. The dead-band booster is designed intentionally with a large dead band (approximately 5 percent of the input span), elastomer seats for tight shutoff, and an adjustable bypass valve connected between the input and output of the booster. The bypass valve is tuned to provide the best compromise between increased actuator stroking speed and positioner/actuator stability.

With the exception of the dead-band booster, the application of booster relays has diminished somewhat by the increased use of current-to-pressure transducers, electropneumatic positioners, and electronic control systems. Transducers and valve positioners serve much the same functionality as the booster relay in addition to interfacing with the electronic process controller.

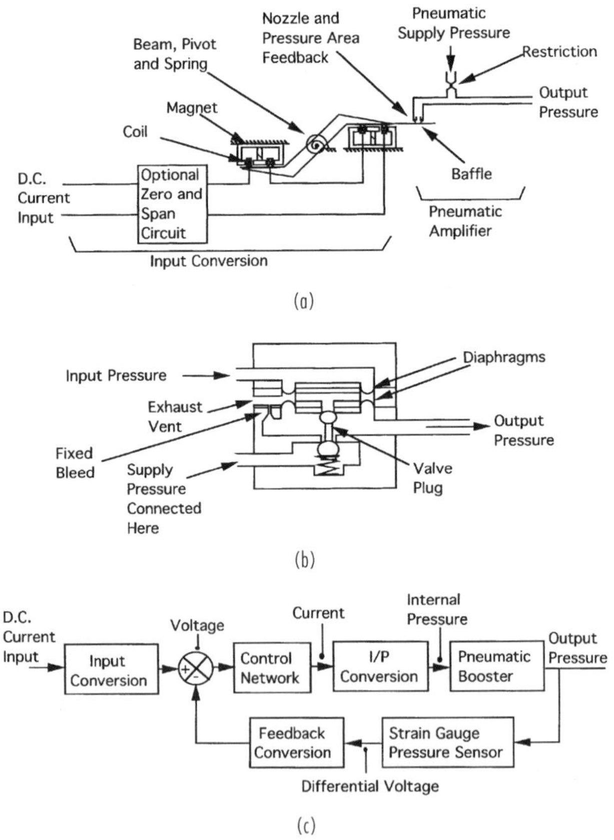

FIG. 8-87 Current-to-pressure transducer component parts. (*a*) Direct-current–pressure conversion. (*b*) Pneumatic booster amplifier (relay). (*c*) Block diagram of a modern I/P transducer.

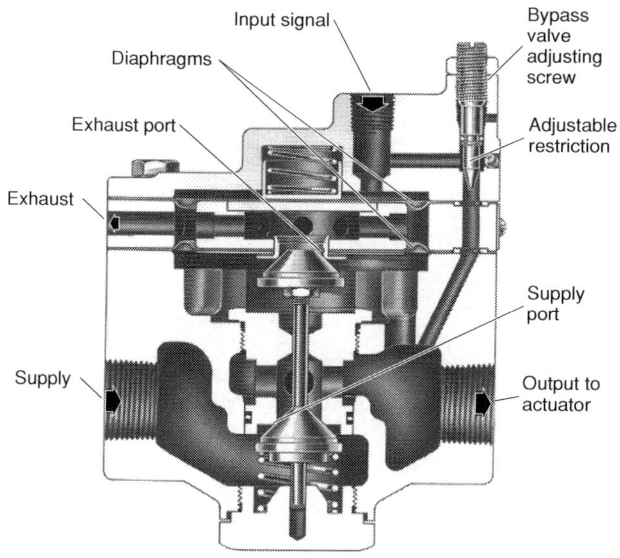

FIG. 8-88 Dead-band booster relay. (*Courtesy Fisher Controls International LLC.*)

Solenoid Valves The electric solenoid valve has two output states. When sufficient electric current is supplied to the coil, an internal armature moves against a spring to an extreme position. This motion causes an attached pneumatic or hydraulic valve to operate. When current is removed, the spring returns the armature and the attached solenoid valve to the deenergized position. An intermediate pilot stage is sometimes used when additional force is required to operate the main solenoid valve. Generally, solenoid valves are used to pressurize or vent the actuator casing for on/off control valve applications and safety shutdown applications.

Trip Valves The trip valve is part of a system used where a specific valve action (i.e., fail up, fail down, or lock in last position) is required when pneumatic supply pressure to the control valve falls below a preset level. Trip systems are used primarily on springless piston actuators requiring fail-open or fail-closed action. An air storage or "volume" tank and a check valve are used with the trip valve to provide power to stroke the valve when supply pressure is lost. Trip valves are designed with hysteresis around the trip point to avoid instability when the trip pressure and the reset pressure settings are too close to the same value.

Limit Switches and Stem Position Transmitters Travel limit switches, position switches, and valve position transmitters are devices that detect the component's relative position, when mounted on the valve, actuator, damper, louver, or other throttling element. The switches are used to operate alarms, signal lights, relays, solenoid valves, or discrete inputs into the control system. The valve position transmitter generates a 4- to 20-mA output that is proportional to the position of the valve.

FIRE AND EXPLOSION PROTECTION

Electrical equipment and wiring methods can be sources of ignition in environments with combustible concentrations of gas, liquid, dust, fibers, or flyings. Most of the time it is possible to locate the electronic equipment away from these hazardous areas. However, where electric or electronic valve-mounted instruments must be used in areas where there is a hazard of fire or explosion, the equipment and installation must meet requirements for safety. Articles 500 through 504 of the National Electrical Code cover the definitions and requirements for electrical and electronic equipment used in the class I (flammable gases or vapors), divisions 1 and 2; class II (combustible dust), divisions 1 and 2; and class III (ignitable fibers or flyings), divisions 1 and 2. Division 1 locations are locations with hazardous concentrations of gases, vapors, or combustible dust under normal operating conditions; hazardous concentration of gases, vapors, or combustible dust that occur frequently due to repair, maintenance, or leakage; or hazardous due to the presence of easily ignitable fibers or materials producing combustible flyings during handling, manufacturing, or use. Division 2 locations are locations that normally do not have ignitable concentrations of gases, vapors, or combustible dust. Division 2 locations might become hazardous through failure of ventilating equipment; adjacent proximity to a class I, division 1 location where ignitable concentrations of gases or vapors might occasionally exist; through dust accumulations on or in the vicinity of the electrical equipment sufficient to interfere with the safe dissipation of heat or by abnormal operation or failure of electrical equipment; or when easily ignitable fibers are stored or handled other than in the process of manufacture. An alternate method used for class I hazardous locations is the European "zone" method described in IEC 60079-10, "Electrical Apparatus for Explosive Gas Atmospheres." The zone designation for class I locations has been adapted by the NEC as an alternate method and is defined in Article 505 of the NEC.

Acceptable protection techniques for electrical and electronic valve accessories used in specific class and division locations include explosion-proof enclosures; intrinsically safe circuits; nonincendive circuits, equipment, and components; dust-ignition-proof enclosures; dusttight enclosures; purged and pressurized enclosures; oil immersion for current-interrupting contacts; and hermetically sealed equipment. Details of these techniques can be found in the *National Electrical Code Handbook*, available from the *National Fire Protection Association*.

Certified testing and approval for control valve devices used in hazardous locations is normally procured by the manufacturer of the device. The manufacturer typically goes to a third-party laboratory for testing and certification. Applicable approval standards are available from CSA, CENELEC, FM, SAA, and UL.

Environmental Enclosures Enclosures for valve accessories are sometimes required to provide protection from specific environmental conditions. The National Electrical Manufacturers Association (NEMA) provides descriptions and test methods for equipment used in specific environmental conditions in NEMA 250. IEC 60529, "Degrees of Protection Provided by Enclosures (IP Code)," describes the European system for classifying the degrees of protection provided by the enclosures of electrical equipment. Rain, windblown dust, hose-directed water, and external ice formation are examples of environmental conditions that are covered by these enclosure standards.

Of growing importance is the electronic control valve device's level of immunity to, and emission of, electromagnetic interference in the chemical valve environment. Electromagnetic compatibility (EMC) for control valve devices is presently mandatory in the European Community and is specified in International Electrotechnical Commission (IEC) 61326, "Electrical Equipment for Measurement Control and Laboratory Use—EMC Requirements." Test methods for EMC testing are found in the series IEC 61000-4, "EMC Compatibility (EMC), Testing and Measurement Techniques." Somewhat more stringent EMC guidelines are found in the German document NAMUR NE21, "Electromagnetic Compatibility of Industrial Process and Laboratory Control Equipment."

ADJUSTABLE-SPEED PUMPS

An alternative to throttling a process with a process control valve and a fixed-speed pump is by adjusting the speed of the process pump and not using a throttling control valve at all. Pump speed can be varied by using variable-speed prime movers such as turbines, motors with magnetic or hydraulic couplings, and electric motors. Each of these methods of modulating pump speed has its own strengths and weaknesses, but all offer energy savings and dynamic performance advantages over throttling with a control valve.

The centrifugal pump directly driven by a variable-speed electric motor is the most commonly used hardware combination for adjustable-speed pumping. The motor is operated by an electronic motor speed controller whose function is to generate the voltage or current waveform required by the motor to make the speed of the motor track the input command signal from the process controller.

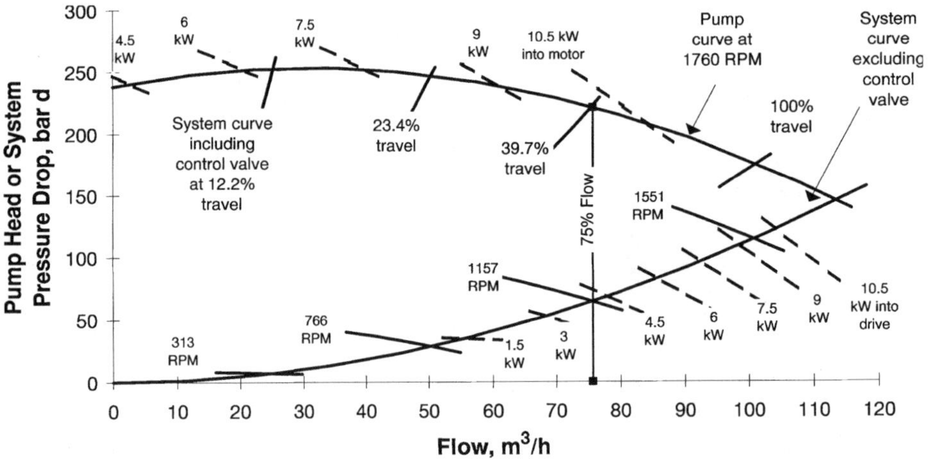

FIG. 8-89 Pressure, flow, and power for throttling a process using a control valve and a constant-speed pump compared to throttling with an adjustable-speed pump.

The most popular form of motor speed control for adjustable-speed pumping is the voltage-controlled pulse-width-modulated (PWM) frequency synthesizer and ac squirrel-cage induction motor combination. The flexibility of application of the PWM motor drive and its 90+ percent electrical efficiency along with the proven ruggedness of the traditional ac induction motor makes this combination popular.

From an energy consumption standpoint, the power required to maintain steady process flow with an adjustable-speed-pump system (three-phase PWM drive and a squirrel-cage induction motor driving a centrifugal pump on water) is less than that required with a conventional control valve and a fixed-speed pump. Figure 8-89 shows this to be the case for a system where 100 percent of the pressure loss is due to flow velocity losses. At 75 percent flow, the figure shows that using the constant-speed pump/control valve results in a 10.1-kW rate, while throttling with the adjustable-speed pump and not using a control valve results in a 4.1-kW rate. This trend of reduced energy consumption is true for the entire range of flows, although amounts vary.

From a dynamic response standpoint, the electronic adjustable-speed pump has a dynamic characteristic that is more suitable in process control applications than those characteristics of control valves. The small amplitude response of an adjustable-speed pump does not contain the dead band or the dead time commonly found in the small amplitude response of the control valve. Nonlinearities associated with friction in the valve and discontinuities in the pneumatic portion of the control valve instrumentation are not present with electronic variable-speed drive technology. As a result, process control with the adjustable-speed pump does not exhibit limit cycles, problems related to low controller gain, and generally degraded process loop performance caused by control valve nonlinearities.

Unlike the control valve, the centrifugal pump has poor or nonexistent shutoff capability. A flow check valve or an automated on/off valve may be required to achieve shutoff requirements. This requirement may be met by automating an existing isolation valve in retrofit applications.

REGULATORS

A regulator is a compact device that maintains the process variable at a specific value in spite of disturbances in load flow. It combines the functions of the measurement sensor, controller, and final control element into one self-contained device. Regulators are available to control pressure, differential pressure, temperature, flow, liquid level, and other basic process variables. They are used to control the differential across a filter press, heat exchanger, or orifice plate. Regulators are used for monitoring pressure variables for redundancy, flow check, and liquid surge relief.

Regulators may be used in gas blanketing systems to maintain a protective environment above any liquid stored in a tank or vessel as the liquid is pumped out. When the temperature of the vessel is suddenly cooled, the regulator maintains the tank pressure and protects the walls of the tank from possible collapse. Regulators are known for their fast dynamic response. The absence of time delay that often comes with more sophisticated control systems makes the regulator useful in applications requiring fast corrective action.

Regulators are designed to operate on the process pressures in the pipeline without any other sources of energy. Upstream and downstream pressures are used to supply and exhaust the regulator. Exhausting is connected back to the downstream piping so that no contamination or leakage to the external environment occurs. This makes regulators useful in remote locations where power is not available or where external venting is not allowed.

The regulator is limited to operating on processes with clean, nonslurry process fluids. The small orifice and valve assemblies contained in the regulator can plug and malfunction if the process fluid that operates the regulator is not sufficiently clean.

Regulators are normally not suited to systems that require constant set-point adjustment. Although regulators are available with capability to respond to remote set-point adjustment, this feature adds complexity to the regulator and may be better addressed by a control-valve-based system. In the simplest of regulators, tuning of the regulator for best control is accomplished by changing a spring, an orifice, or a nozzle.

Self-Operated Regulators Self-operated regulators are the simplest form of regulator. This regulator (see Fig. 8-90*a*) is composed of a main throttling valve, a diaphragm or piston to sense pressure, and a spring. The self-contained regulator is completely operated by the process fluid, and no outside control lines or pilot stage is used. In general, self-operated regulators are simple in construction, are easy to operate and maintain, and are usually stable devices. Except for some of the pitot-tube types, self-operated regulators have very good dynamic response characteristics. This is so because any change in the controlled variable registers directly and immediately upon the main diaphragm to produce a quick response to the disturbance.

The disadvantage of the self-operated regulator is that it is not generally capable of maintaining a set point as load flow is increased. Because of the proportional nature of the spring and diaphragm-throttling effect, offset from set point occurs in the controlled variable as flow increases. Figure 8-91 shows a typical regulation curve for the self-contained regulator.

Reduced set-point offset with increasing load flow can be achieved by adding a pitot tube to the self-operated regulator. The

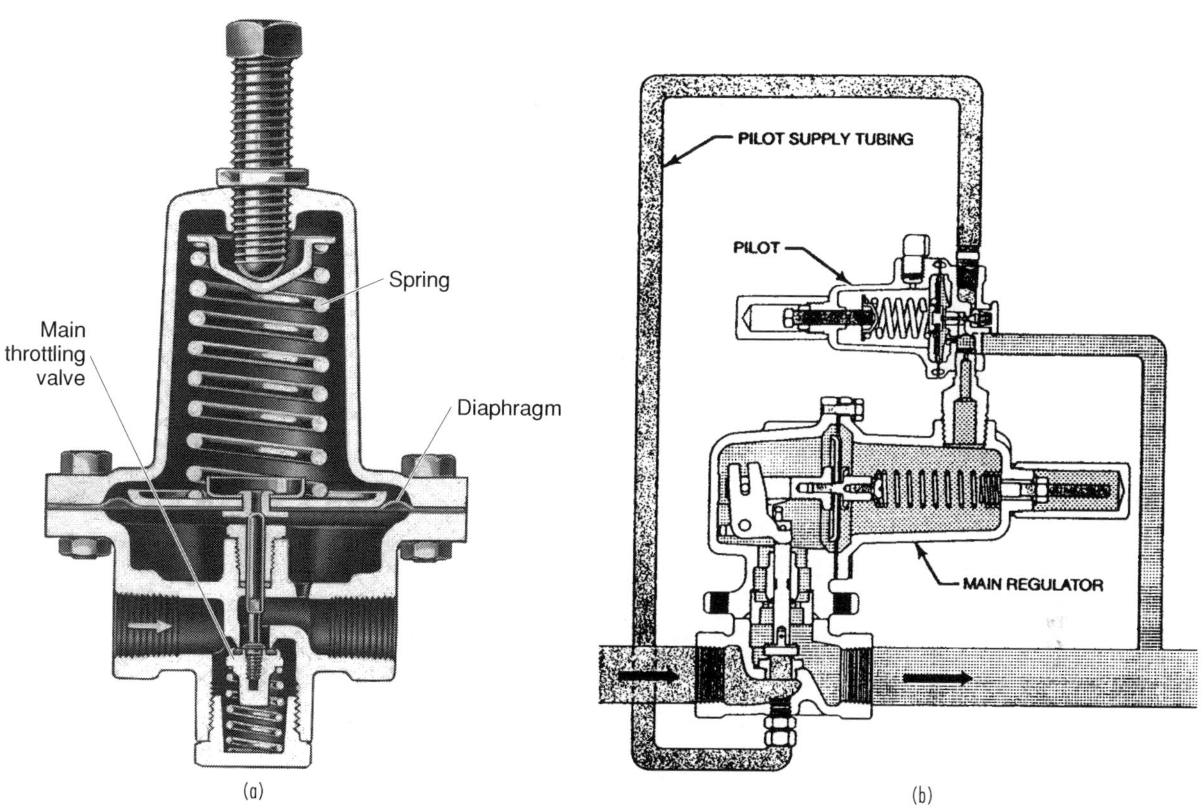

FIG. 8-90 Regulators. (*a*) Self-operated. (*b*) Pilot-operated. (*Courtesy Fisher Controls International LLC.*)

tube is positioned somewhere near the vena contracta of the main regulator valve. As flow though the valve increases, the measured feedback pressure from the pitot tube drops below the control pressure. This causes the main valve to open or boost more than it would if the static value of control pressure were acting on the diaphragm. The resultant effect keeps the control pressure closer to the set point and thus prevents a large drop in process pressure during high-load-flow conditions. Figure 8-91 shows the improvement that the pitot-tube regulator provides over the regulator without the tube. A side effect of adding a pitot-tube method is that the response of the regulator can be slowed due to the restriction provided by the pitot tube.

Pilot-Operated Regulators Another category of regulators uses a pilot stage to provide the load pressure on the main diaphragm. This pilot is a regulator itself that has the ability to multiply a small change in downstream pressure into a large change in pressure applied to the regulator diaphragm. Due to this high-gain feature, pilot-operated

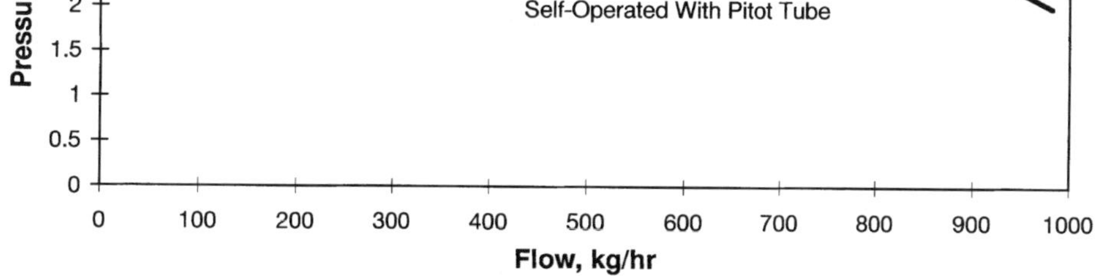

FIG. 8-91 Pressure regulation curves for three regulator types.

regulators can achieve a dramatic improvement in steady-state accuracy over that achieved with a self-operated regulator. Figure 8-91 shows for regulation at high flows the pilot-operated regulator is the best of the three regulators shown.

The main limitation of the pilot-operated regulator is stability. When the gain in the pilot amplifier is raised too much, the loop can become unstable and oscillate or hunt. The two-path pilot regulator (see Fig. 8-90*b*) is also available. This regulator combines the effects of self-operated and the pilot-operated styles and mathematically produces the equivalent of proportional plus reset control of the process pressure.

Overpressure Protection Figure 8-91 shows a characteristic rise in control pressure that occurs at low or zero flow. This lockup tail is due to the effects of imperfect plug and seat alignment and the elastomeric effects of the main throttle valve. If, for some reason, the main throttle valve fails to completely shut off, or if the valve shuts off but the control pressure continues to rise for other reasons, the lockup tail could get very large, and the control pressure could rise to extremely high values. Damage to the regulator or the downstream pressure volume could occur.

To avoid this situation, some regulators are designed with a built-in overpressure relief mechanism. Overpressure relief circuits usually are composed of a spring-opposed diaphragm and valve assembly that vents the downstream piping when the control pressure rises above the set-point pressure.

PROCESS CONTROL AND PLANT SAFETY

GENERAL REFERENCE: *Guidelines for Safe Automation of Chemical Processes,* AIChE Center for Chemical Process Safety, New York, 1993.

Accidents in chemical plants make headline news, especially when there is loss of life or the general public is affected in even the slightest way. This increases the public's concern and may lead to government action. The terms *hazard* and *risk* are defined as follows:

- *Hazard.* A potential source of harm to people, property, or the environment.
- *Risk.* Possibility of injury, loss, or an environmental accident created by a hazard.

Safety is the freedom from hazards and thus the absence of any associated risks. Unfortunately, absolute safety cannot be realized.

The design and implementation of safety systems must be undertaken with a view to two issues:

- *Regulatory.* The safety system must be consistent with all applicable codes and standards as well as "generally accepted good engineering practices."
- *Technical.* Just meeting all applicable regulations and "following the crowd" do not relieve a company of its responsibilities. The safety system must work.

The regulatory environment will continue to change. As of this writing, the key regulatory instrument is OSHA 29 CFR 1910.119, "Process Safety Management of Highly Hazardous Chemicals," which pertains to process safety management within plants in which certain chemicals are present.

In addition to government regulation, industry groups and professional societies are producing documents ranging from standards to guidelines. Two applicable standards are IEC 61508, "Functional Safety of Electrical/Electronic/Programmable Electronic Safety-related Systems," and ANSI/ISA S84.01, "Application of Safety Instrumented Systems for the Process Industries." *Guidelines for Safe Automation of Chemical Processes* from the American Institute of Chemical Engineers' Center for Chemical Process Safety (1993) provides comprehensive coverage of the various aspects of safety; and although short on specifics, it is very useful to operating companies developing their own specific safety practices (i.e., it does not tell you what to do, but it helps you decide what is proper for your plant).

The ultimate responsibility for safety rests with the operating company; OSHA 1910.119 is clear on this. Each company is expected to develop (and enforce) its own practices in the design, installation, testing, and maintenance of safety systems. Fortunately, some companies make these documents public. Monsanto's *Safety System Design Practices* was published in its entirety in the proceedings of the International Symposium and Workshop on Safe Chemical Process Automation, Houston, Texas, September 27–29, 1994 (available from the American Institute of Chemical Engineers' Center for Chemical Process Safety).

ROLE OF AUTOMATION IN PLANT SAFETY

As microprocessor-based controls displaced hardwired electronic and pneumatic controls, the impact on plant safety has definitely been positive. When automated procedures replace manual procedures for routine operations, the probability of human errors leading to hazardous situations is lowered. The enhanced capability for presenting information to the process operators in a timely manner and in the most meaningful form increases the operator's awareness of current conditions in the process. Process operators are expected to exercise due diligence in the supervision of the process, and timely recognition of an abnormal situation reduces the likelihood that the situation will progress to the hazardous state. Figure 8-92 depicts the layers of safety protection in a typical chemical plant. Although microprocessor-based process controls enhance plant safety, their primary objective is efficient process operation. Manual operations are automated to reduce variability, to minimize the time required, to increase productivity, and so on. Remaining competitive in the world market demands that the plant be operated in the best manner possible, and microprocessor-based process controls provide numerous functions that make this possible. Safety is never compromised in the effort to increase competitiveness, but enhanced safety is a by-product of the process control function and is not a primary objective. By attempting to maintain process conditions at or near their design values, the process controls also attempt to prevent abnormal conditions from developing within the process.

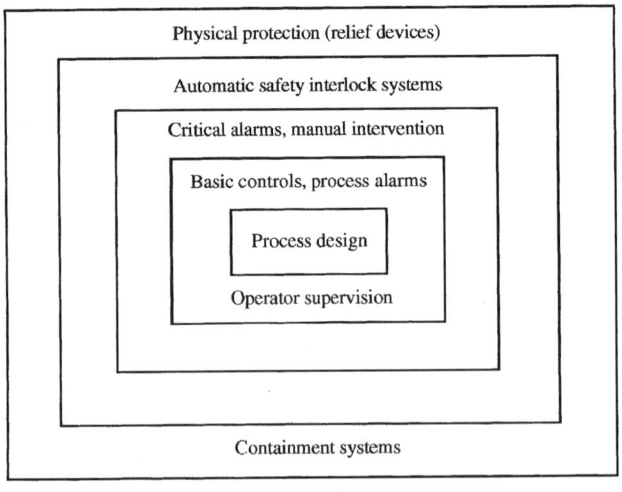

FIG. 8-92 Layers of safety protection in chemical plants.

Although process controls can be viewed as a protective layer, this is really a by-product and not the primary function. Where the objective of a function is specifically to reduce risk, the implementation is normally not within the process controls. Instead, the implementation is within a separate system specifically provided to reduce risk. This system is generally referred to as the safety interlock system.

As safety begins with the process design, an inherently safe process is the objective of modern plant designs. When this cannot be achieved, process hazards of varying severity will exist. Where these hazards put plant workers and/or the general public at risk, some form of protective system is required. Process safety management addresses the various issues, ranging from assessment of the process hazard to ensuring the integrity of the protective equipment installed to cope with the hazard. When the protective system is an automatic action, it is incorporated into the safety interlock system, not within the process controls.

INTEGRITY OF PROCESS CONTROL SYSTEMS

Ensuring the integrity of process controls involves hardware issues, software issues, and human issues. Of these, the hardware issues are usually the easiest to assess and the software issues the most difficult.

The hardware issues are addressed by providing various degrees of redundancy, by providing multiple sources of power and/or an uninterruptible power supply, and the like. The manufacturers of process controls provide a variety of configuration options. Where the process is inherently safe and infrequent shutdowns can be tolerated, nonredundant configurations are acceptable. For more-demanding situations, an appropriate requirement might be that no single component failure be able to render the process control system inoperable. For the very critical situations, triple-redundant controls with voting logic might be appropriate. The difficulty lies in assessing what is required for a given process.

Another difficulty lies in assessing the potential for human errors. If redundancy is accompanied with increased complexity, the resulting increased potential for human errors must be taken into consideration. Redundant systems require maintenance procedures that can correct problems in one part of the system while the remainder of the system is in full operation. When maintenance is conducted in such situations, the consequences of human errors can be rather unpleasant.

The use of programmable systems for process control presents some possibilities for failures that do not exist in hardwired electromechanical implementations. Probably of greatest concern are latent defects or "bugs" in the software, either the software provided by the supplier or the software developed by the user. The source of this problem is very simple. There is no methodology available that can be applied to obtain absolute assurance that a given set of software is completely free of defects. Increased confidence in a set of software is achieved via extensive testing, but no amount of testing results in absolute assurance that there are no defects. This is especially true of real-time systems, where the software can easily be exposed to a sequence of events that was not anticipated. Just because the software performs correctly for each event individually does not mean that it will perform correctly when two (or more) events occur at nearly the same time. This is further complicated by the fact that the defect may not be in the programming; it may be in how the software was designed to respond to the events.

The testing of any collection of software is made more difficult as the complexity of the software increases. Software for process control has become progressively complex, mainly because the requirements have become progressively demanding. To remain competitive in the world market, processes must be operated at higher production rates, within narrower operating ranges, closer to equipment limits, and so on. Demanding applications require sophisticated control strategies, which translate to more-complex software. Even with the best efforts of both supplier and user, complex software systems are unlikely to be completely free of defects.

CONSIDERATIONS IN IMPLEMENTATION OF SAFETY INTERLOCK SYSTEMS

Where hazardous conditions can develop within a process, a protective system of some type must be provided. Sometimes this is in the form of process hardware such as pressure relief devices. However, sometimes logic must be provided for the specific purpose of taking the process to a state where the hazardous condition cannot exist. The term *safety interlock system* is normally used to designate such logic.

The purpose of the logic within the safety interlock system is very different from that of the logic within the process controls. Fortunately, the logic within the safety interlock system is normally much simpler than the logic within the process controls. This simplicity means that a hardwired implementation of the safety interlock system is usually an option. Should a programmable implementation be chosen, this simplicity means that latent defects in the software are less likely to be present. Most safety systems only have to do simple things, but they must do them very, very well.

The difference in the nature of process controls and safety interlock systems leads to the conclusion that these two should be physically separated (see Fig. 8-92). That is, safety interlocks should not be piggybacked onto a process control system. Instead, the safety interlocks should be provided by equipment, either hardwired or programmable, that is dedicated to the safety functions. As the process controls become more complex, faults are more likely. Separation means that faults within the process controls have no consequences in the safety interlock system.

Modifications to the process controls are more frequent than modifications to the safety interlock system. Therefore, physically separating the safety interlock system from the process controls provides the following benefits:

1. The possibility of a change to the process controls leading to an unintentional change to the safety interlock system is eliminated.
2. The possibility of a human error in the maintenance of the process controls having consequences for the safety interlock system is eliminated.
3. Management of change is simplified.
4. Administrative procedures for software version control are more manageable.

Separation also applies to the measurement devices and actuators. Although the traditional point of reference for safety interlock systems is a hardwired implementation, a programmed implementation is an alternative. The potential for latent defects in software implementation is a definite concern. Another concern is that solid-state components are not guaranteed to fail to the safe state. The former is addressed by extensive testing; the latter is addressed by manufacturer-supplied and/or user-supplied diagnostics that are routinely executed by the processor within the safety interlock system. Although issues must be addressed in programmable implementations, the hardwired implementations are not perfect either.

Where a programmed implementation is deemed to be acceptable, the choice is usually a programmable logic controller that is dedicated to the safety function. PLCs are programmed with the traditional relay ladder diagrams used for hardwired implementations. The facilities for developing, testing, and troubleshooting PLCs are excellent. However, for PLCs used in safety interlock systems, administrative procedures must be developed and implemented to address the following issues:

1. Version controls for the PLC program must be implemented and rigidly enforced. Revisions to the program must be reviewed in detail and thoroughly tested before implemention in the PLC. The various versions must be clearly identified so that there can be no doubt as to what logic is provided by each version of the program.
2. The version of the program that is currently being executed by the PLC must be known with absolute certainty. It must be impossible for a revised version of the program undergoing testing to be downloaded to the PLC.

Constant vigilance is required to prevent lapses in such administrative procedures.

INTERLOCKS

An interlock is a protective response initiated on the detection of a process hazard. The interlock system consists of the measurement devices, logic solvers, and final control elements that recognize the hazard and initiate an appropriate response. Most interlocks consist of one or more logic conditions that detect out-of-limit process conditions and respond by driving the final control elements to the safe states. For example, one must specify that a valve fails open or fails closed.

The potential that the logic within the interlock could contain a defect or bug is a strong incentive to keep it simple. Within process plants, most interlocks are implemented with discrete logic, which means either hardwired electromechanical devices or programmable logic controllers.

The discrete logic within process plants can be broadly classified as follows:

1. *Safety interlocks.* These are designed to protect the public, the plant personnel, and possibly the plant equipment from process hazards. These are implemented within the safety interlock system.

2. *Process actions.* These are designed to prevent process conditions that would unduly stress equipment (perhaps leading to minor damage), lead to off-specification product, and so on. Basically, the process actions address hazards whose consequences essentially lead to a monetary loss, possibly even a short plant shutdown. Although sometimes referred to as interlocks, process actions address situations that are not deemed to be process hazards.

Implementation of process actions within process control systems is perfectly acceptable. Furthermore, it is also permissible (and probably advisable) for responsible operations personnel to be authorized to bypass or ignore a process action. Safety interlocks must be implemented within the separate safety interlock system. Bypassing or ignoring safety interlocks by operations personnel is simply not permitted. When this is necessary for actions such as verifying that the interlock continues to be functional, such situations must be infrequent and incorporated into the design of the interlock.

Safety interlocks are assigned to categories that reflect the severity of the consequences, should the interlock fail to perform as intended. The specific categories used within a company are completely at the discretion of the company. However, most companies use categories that distinguish among the following:

1. *Hazards that pose a risk to the public.* Complete redundancy is normally required.

2. *Hazards that could lead to injury of company personnel.* Partial redundancy is often required (e.g., redundant measurements but not redundant logic).

3. *Hazards that could result in major equipment damage and consequently lengthy plant downtime.* No redundancy is normally required for these, although redundancy is always an option.

Situations resulting in minor equipment damage that can be quickly repaired do not generally require a safety interlock; however, a process action might be appropriate.

A process hazards analysis is intended to identify the safety interlocks required for a process and to provide the following for each:

1. The hazard that is to be addressed by the safety interlock
2. The classification of the safety interlock
3. The logic for the safety interlock, including inputs from measurement devices and outputs to actuators

The process hazards analysis is conducted by an experienced, multidisciplinary team that examines the process design, plant equipment, operating procedures, and so on, using techniques such as hazard and operability studies (HAZOP), failure mode and effect analysis (FMEA), and others. The process hazards analysis recommends appropriate measures to reduce the risk, including (but not limited to) the safety interlocks to be implemented in the safety interlock system.

Diversity is recognized as a useful approach to reduce the number of defects. The team that conducts the process hazards analysis does not implement the safety interlocks but provides the specifications for the safety interlocks to another organization for implementation. This organization reviews the specifications for each safety interlock, seeking clarifications as necessary from the process hazards analysis team and bringing any perceived deficiencies to the attention of the process hazards analysis team.

Diversity can be used to further advantage in redundant configurations. Where redundant measurement devices are required, different technology can be used for each. Where redundant logic is required, one can be programmed and one hardwired. Reliability of the interlock systems has two aspects:

1. It must react, should the hazard arise.
2. It must not react when there is no hazard.

Emergency shutdowns often pose risks in themselves, and therefore they should be undertaken only when truly appropriate. The need to avoid extraneous shutdowns is not motivated by a desire simply to avoid disruption in production operations.

Although safety interlocks can inappropriately initiate shutdowns, the process actions are usually the major source of problems. It is possible to configure so many process actions that it is not possible to operate the plant.

TESTING

As part of the detailed design of each safety interlock, written test procedures must be developed for the following purposes:

1. Ensure that the initial implementation complies with the requirements defined by the process hazards analysis team.

2. Ensure that the interlock (hardware, software, and I/O) continues to function as designed. The design must also determine the time interval over which this must be done. Often these tests must be done with the plant in full operation.

The former is the responsibility of the implementation team and is required for the initial implementation and following any modification to the interlock. The latter is the responsibility of plant maintenance, with plant management responsible for seeing that it is done at the specified interval of time.

Execution of each test must be documented, showing when it was done, by whom, and the results. Failures must be analyzed for possible changes in the design or implementation of the interlock. These tests must encompass the complete interlock system, from the measurement devices through the final control elements. Merely simulating inputs and checking the outputs is not sufficient. The tests must duplicate the process conditions and operating environments as closely as possible. The measurement devices and final control elements are exposed to process and ambient conditions and thus are usually the most likely to fail. Valves that remain in the same position for extended periods may stick in that position and not operate when needed. The easiest component to test is the logic; however, this is the least likely to fail.

Made in United States
Orlando, FL
04 November 2026

90200180R00055